FAT PLANET

The Obesity Trap and How We Can Escape It

DR DAVID LEWIS & DR MARGARET LEITCH

arrow books

1 3 5 7 9 10 8 6 4 2

Arrow Books
20 Vauxhall Bridge Road
London SW1V 2SA

Arrow Books is part of the Penguin Random House
group of companies whose addresses can be found at
global.penguinrandomhouse.com

Penguin
Random House
UK

First published by Random House Books in 2015
First published in paperback by Arrow Books in 2016

www.randomhouse.co.uk

A CIP catalogue record for this book is available
from the British Library

ISBN 9780099594123

Typeset by Palimpsest Book Production Ltd, Falkirk, Stirlingshire

Printed and bound by CPI Group (UK) Ltd, Croydon CR0 4YY

Penguin Random House is committed to a sustainable future for
our business, our readers and our planet. This book is made from
Forest Stewardship Council® certified paper.

MIX
Paper from
responsible sources
FSC® C018179

FAT
PLANET

Dr David Lewis is a neuropsychologist, director of research at Mindlab International, and the author of more than two dozen bestselling books. He is a Chartered Member of the British Psychological Society, a Fellow of the International Stress Management Association, and founder of Action on Phobias and Stresswatch; a non-profit charity advising companies on stress management. He has also been the on-screen consultant to Channel 4's acclaimed *Secret Eaters* diet programme.

Dr Margaret Leitch is a research scientist with PhD, MSc, and Honours BA degrees in the field of Experimental Psychology, Research Methods and Statistics, and Honours Psychology. She has conducted research into the psychological and neuroscientific underpinnings of obesity, including three years of longitudinal patient monitoring. She is a specialist in identifying the predictive signs that express a vulnerability for developing overeating habits. She is co-author of the bestselling book *Lose It Right*.

Praise for *Fat Planet*

'*Fat Planet*'s authors are distinguished researchers who are masters at describing the latest scientific insights in a simple and engaging way. A must read.'

Professor Pierre Chandon, Director,
Sorbonne INSEAD Behavioural Lab

'*Fat Planet* is a spot-on analysis of the causes of the obesity epidemic, full of fascinating and surprising research evidence demonstrating that most of us can neither recognize nor resist the forces that lead us to eat too much. A recommended read for anyone who wants to make our food environment safer for all.'

Deborah A. Cohen, Senior Natural Scientist,
RAND Corporation

'A provocative and eye-opening examination of the world's obesity crisis. A must read for those wishing to understand the true causes of obesity and what we can do about it.'

Hank Cardello, Senior Fellow & Director,
Obesity Solutions Initiative, Hudson Institute

We dedicate this book to the millions of men and women around the world who want to understand the real reasons why they are having personal issues with their weight.

Contents

Acknowledgements

We have been tremendously fortunate, whilst writing this book, in having the assistance of some of the world's foremost experts in obesity. This project would not have been possible without the generosity and dedication of these leading neuroscientists, physiologists and policy experts.

Drs Nicole Avena and Alan Geliebter from the University of Columbia, and Kent Berridge, University of Michigan, are internationally respected specialists in neurocognitive aspects of food pleasure and reward. Given the extent to which obesity is driven by the pleasure of consumption, their expertise was invaluable in the completion of this book.

We would also like to extend our gratitude to Dr Emma Boyland at the University of Liverpool and Dr Laura Forbes at the University of Guelph who provided insight into nutrition and food marketing.

Dr Jean-Philippe Chaput from the Children's Hospital of Eastern Ontario, Canada, advised us on the link between circadian rhythms and appetite control.

Dr Jeffrey Friedman, a leading expert in molecular genetics and recipient of *Time Magazine*'s Best in Science, gave us a fascinating personal account of his discovery of the hormone leptin.

Market researcher and psychophysicist Dr Howard Moskowitz, CEO of i-Novation Inc. as well as President of Moskowitz Jacobs Inc., contributed greatly to our understanding of the factors that make food taste delicious.

Dr Sean Kelly, former lecturer in neurotoxicology at the University of Newcastle-upon-Tyne, provided valuable feedback on medical and physiological complications associated with excess adiposity.

We are also most grateful to Rand Corporation policy expert Dr Deborah Cohen and to Dr Yoni Freedhoff, one of Canada's most respected authorities on food policy, whose insights enabled us to structure a clear set of guidelines on this complex and politically sensitive topic.

Hank Cardello, former food industry executive and author of *Stuffed: An Insider's Look at Who's (Really) Making America Fat* offered excellent perspectives on the business case for producing healthier and more nutritionally sound food.

Our grateful thanks to Steven Matthews, LLB, LLM, MA who was kind enough to read the manuscript and provide an astute and perceptive critique of the early drafts; to Matt Briggs for the medical illustrations; to Duncan Smith and Amy Maddock of Mindlab International for their support and advice; to Dan Jones, hypnotherapist, and to Robert E. Lamb PhD of Michigan City, North Dakota for his insights into obesity in the US.

We should also like to acknowledge the assistance of Jane Lee, Southern Railways press officer, for arranging filming facilities, and videographer Jerry Laurence and stills photographer Afonso Palma. We are very grateful to Lee McMurray, Emily Kennedy and Curtis Leighton-Jones, Endemol Series Producers on Series 1, 2 and 3 of Channel 4's *Secret Eaters* and also to Anna Richardson, who presented the series, for enabling us to replicate some key studies into the psychology of obesity.

Last, but certainly not least, we would like to extend our grateful thanks to our Random House editors, Nigel Wilcockson, for commissioning the book and suggesting its title, and Harry Scoble, for his excellent and thorough work on the manuscript.

We sincerely appreciate all your time and support in helping us provide these insights into the world's most serious and most misunderstood health crisis.

Introduction

'We can all agree that . . . all children should have the basic
nutrition they need to learn and grow and to pursue their
dreams, because in the end, nothing is more important than the
health and well-being of our children.'
Michelle Obama[1]

Our planet is fat and getting fatter. We are living in the midst of a
global obesity pandemic, a pandemic which began some thirty years
ago in industrialised countries, before rapidly spreading to the developing world.[2] Today an estimated 1.46 billion adults worldwide are
overweight and 671 million are obese.[3] And it's not just adults who
are affected, of course. In some countries over a quarter of children
are too fat to be healthy, more than double the proportion at the
start of the 1970s.[4] Worldwide, around 170 million children under
the age of eighteen are now either overweight or obese.

In the UK, dubbed 'the Fat Man of Europe' by the Academy of
Royal Colleges, the number of obese adults has risen by 300% over
the past three decades.[5]

In 1966, just under 80% of adults were in the 'healthy' weight
range, with only around one in seven (13%) men being overweight
and one in a hundred (1.2%) obese. Among women, just one in
ten (9%) was overweight or obese. By 1996, those in the healthy
range had declined to 28% of men and 38% of women. At the
same time, the proportion of both overweight and obese people
had risen significantly. Almost half of all men (49%) and a third
of all women (36%) had become overweight, while about a quarter
of both sexes were obese (23% of men and 26% of women).[6]

In the decade and a half since then, these differences have become even greater, with two thirds of adult males and more than half of adult females now either overweight or obese. Half a century ago, the average British woman could wear a UK dress size 12 comfortably. By 2006 she would have needed a size 14. Today, almost half (47%) would struggle to squeeze into anything smaller than a size 16 or even larger. If this trend continues, by 2030 there will be 11 million more obese men and women in the country.[7]

Obesity rates are predicted to rise as rapidly throughout Western Europe as they have in the UK. Laura Webber, from the UK Health Forum, predicts that by 2030, 80% of adults in Spain, the Czech Republic and Poland, and 90% of adults in Ireland will be overweight or obese.[8]

The story in the US is even worse. Over the past three decades, America has become famous for its fatness. Despite having a greater awareness of obesity than almost any other nation, two thirds (68%) of American adults are overweight and a third (34%) obese. Between 1960 and 2004, the proportion of men aged between twenty and forty who fell into these categories rose from 45% to around 70%. In twelve states, more than 30% of the population are obese, and not a single state has an obesity rate lower than 20%.[9] The prediction here is that, by 2030, there will be an additional 65 million obese people in the US.

But despite our tendency to think of obesity as a problem largely confined to wealthier countries, in actual fact there is nowhere where the problem of weight gain is happening faster or more dangerously than in the developing world, where healthy, traditional diets are being replaced with far more energy-dense, less nutritionally sound, Western-style eating habits.[10] The process of globalisation and foreign direct investment into food processing has been directly linked with the shift in diets amongst emerging economies.[11] In India, for example, almost 75% of Foreign Direct Investment goes towards highly processed foods. The shift in diets is referred to as 'nutrition transition', which serves as a euphemism for the disastrous consequences that obesity will spell for the developing world.[12]

China provides an important case study in this regard. It has only been within the last twenty years that the Chinese have had access to a 'Western' diet and the consequences have been disastrous. While obesity was never traditionally a problem in Asia, today there are around 350 million overweight and 60 million obese people in China.[13] This is around one quarter of the population. At the same time, roughly a hundred million are undernourished – a stark contrast, which goes to show just how warped our current food systems are. It has become increasingly common to find obesity coexisting with malnutrition, sometimes even within the same household.[14]

If the pattern set by China and other developing countries continues to be followed, and traditional foods continue to fall out of favour, then obesity and the ill health that accompanies it will become a crippling societal norm.

What is particularly troubling in this general increase in weight we are seeing across the world is how severely children are affected.

In Britain, obesity is rising twice as fast among children as among adults. Today more than one in seven (16%) youngsters aged between 6 and 15 are obese, a percentage that is three times higher than ten years ago. It is conservatively estimated that, within six years, a quarter (23.5%) of boys and a third (32%) of girls will be either overweight or obese. By comparison, only 4% of French children are obese. A third of all obese children in Europe are now British.[15]

In the US, the largest industrialised country on the planet, obesity rates in children are far worse, and are still skyrocketing. Between 2009 and 2010, 32% of children between the ages of 2 and 19 were overweight or obese.[16] If these trends continue, experts warn that by 2030 there will be 65 million more obese adults in the USA. Whether this is a reflection of an inherent biological tendency towards the development of obesity in certain nations, or the consequence of socio-economic conditions, is still up for debate (though the latter seems more likely).

Regardless, with a deluge of processed cereals and fatty, sugary foods now being made available in developing countries, it is hardly surprising to learn that childhood obesity rates are already on the rise globally, with a 60% increase in childhood obesity since 1990.[17] Patterns of weight gain, similar to those present in the developed nations, can now be found among children and teenagers in developing countries. Over the past ten years, the percentage of those who are overweight or obese has risen from around one in ten for both boys (8.1%) and girls (8.4%) to one in eight (12.9% for boys and 13.4% for girls).[18]

This additional weight has a direct impact on the health of the individuals concerned. The 'metabolic syndrome' is a term used to describe a cluster of health problems that arise when people become excessively overweight. These include an increased risk of high blood pressure (hypertension), cancer, cardiovascular disease, dementia, infertility, painful joints, depression, raised blood sugar and cholesterol levels and, above all, Type II (adult onset) diabetes.

In the UK, over three million adults (representing 6% of the population, or 1 person in 17) were diagnosed with diabetes in 2013.[19] In the USA, researchers predict the number of sufferers will have risen from 11 million in 2000 (4.0% of the population) to 29 million in 2050 (7.2% of the population).[20] Many doctors, on both sides of the Atlantic, believe these numbers significantly underestimate the extent of the problem, since there are likely to be many more individuals whose diabetes has yet to be diagnosed.

Claire Wang, from the Department of Health Policy and Management at the Mailman School of Public Health, warns that if these trends continue, within five years there will be 'an additional 6–8.5 million cases of diabetes, 5.7–7.3 million cases of heart disease and stroke, 492,000–669,000 additional cases of cancer and

26–55 million quality-adjusted life years forgone in both the USA and UK combined.'[21]

Another recently identified health problem affecting overweight or obese women is an increased likelihood of developing rheumatoid arthritis later in life. According to Bing Lu and his colleagues, from Harvard University's medical and public health schools in Boston, the risk of developing the condition is a third higher (35%) for women who are overweight at the age of eighteen or older. Furthermore, the chances of their developing a more severe form of rheumatoid arthritis, known as seropositive RA, increases by almost 50%.[22]

Obesity is not only linked to a decline in physical health, but also in mental health; the association between obesity and depression is well established. Yet, there may be an even more worrying consequence of excessive weight. Excess adipose tissue is now also known to cause inflammation in key areas in the brain directly related to appetite control, including the hypothalamus.[23] It's not just physiological dysfunction in the form of heart disease or cancer that is associated with excess weight; rather, we are beginning to see a pattern of cognitive decline that is associated with obesity. For example, we now know obesity presents a major risk factor for the development of Alzheimer's Disease, a condition doctors are now calling Type III diabetes.[24]

The overall picture is clear: the whole world, rich and poor, young and old, is gradually getting fatter. The consequences of this, both for individual sufferers, and at a wider societal level, will be severe. If we continue eating ourselves sick, billions will suffer ill health and shortened lives. The financial and social costs will become unsustainable and our planet uninhabitable. While there can be no easy answers to what is – as we will show – a complex and multifaceted problem, there are certain practical steps that individuals, companies and governments can and must take to bring the obesity pandemic under control.

These are the issues which we will address in this book.

PART ONE

The Obesity Blame Game

'This is what people don't understand: obesity is . . . not a lifestyle choice where people are just eating and not exercising. It's because we are getting empty calories – lots of calories – but no nutrition.'[1]
American celebrity chef, Tom Colicchio[2]

CHAPTER 1

What It Feels Like to Be Fat

'Did this ever happen to you? You were walking in the mall –
either by yourself or with a friend and you caught sight of
yourself in a mirror . . . and found yourself wondering, "*Who is
that and how did I get like this?*"'
Diane Carbonell[1]

'Look at that fat lump,' we heard a young male student mutter to
his friends. 'Isn't she disgusting!'

Most people would be deeply offended by such a spiteful remark.
We were delighted!

The success or failure of the study we were about to undertake
depended on our audience reacting to us realistically. This appalling
behaviour only served to demonstrate 'fatism' is as real as it is
cruel. Our area of scientific research lies in the psychological and
neurocognitive aspects of weight gain. We study the factors that
influence the food people buy and the amount they eat, factors
which are often subtle or hidden. But while we understood obesity[2]
from a scientific point of view, there was one aspect of the issue
about which we were entirely ignorant.

What it feels like to be fat.

Today we were going to remedy, in some small and limited way,
that gap in our knowledge. One of us (Margaret) had been transformed from a slim and athletic young woman into someone not
just a little overweight but seriously obese. Margaret had donned
a kapok-lined fat suit. She had done so previously, while working
at the University of Alberta's Diabetes Research Institute. That had

been just for a short period to see what it felt like. And today, rather than staying in a research institute where people with a Body Mass Index (BMI) over 50 are a familiar sight, she would be out on the streets among strangers.

With two major universities and a large gay community, the seaside city of Brighton is a lively, vibrant and youthful place where, for many, fitness and self-image is important. We thought that people there would likely be critical, perhaps openly, of a seriously obese person. Margaret was going to spend a day there.

Why were we conducting this study?

Certainly not to mock or demean those with a genuine weight problem. Nor to trivialise the very real difficulties they experience on a daily basis. Rather, we hoped to gain some first-hand insights into the physical and mental challenges of being obese.

To ensure an accurate record of everything that happened, Margaret was 'wired for sound' with a hidden microphone that would record everything other people said, as well as enabling her to keep up a commentary of her thoughts, feelings and experiences. She would also be followed, at a discreet distance, by a videographer, whose task it was to record the way other people responded to her in the street, on public transport and while shopping. Here's what happened on what turned out to be one of the most tiring and dispiriting days of her life.

In the Mindlab Laboratory: 8.50 a.m.

Struggling to pull on my supersized jeans. I am starting to appreciate the very real practical difficulties of being obese. I knew before I started that adjusting mentally and physically to my newly supersized body would be tricky. But, even taking my inexperience as an overweight woman into account, I am stunned at how hard it is to dress. Hoisting those jeans over my vastly expanded legs is proving far harder and more time consuming than I expected. I can't believe how little give they have in them.

9.25 a.m.

It has taken me over twenty minutes to pull up my jeans and, after an exhausting struggle, fasten the belt. While slipping into my loose fitting top isn't overly challenging, putting on my shoes proves frustratingly hard and doing up the laces even more so. Perhaps I should have bought some with Velcro fasteners.

9.55 a.m.

Finally dressed. No time to bother with hair or make-up but for once I don't care. I am reluctant to see myself in the mirror. It isn't that I feel ashamed but rather that some actions suddenly seem less important than others. I find myself adopting a kind of utilitarian perspective on my priorities. Being concerned with my appearance feels frivolous and unnecessary when just getting my clothes on has proved such a major undertaking.

Perhaps this helps explain why some obese people can struggle to see themselves as overweight; many avoid mirrors to shield themselves from the physical reality of their bodies.[3] Overweight people with whom I have discussed this confirm it's a psychological defence mechanism they sometimes use.

'I've denied that the image I see in the mirror is really me, even though in my heart, I know it is,' wrote Elizabeth Hawksworth in her moving blog entitled *What It's Really Like to Live as a Fat Person Every Day*. She continued: 'I have denied that it's a problem for me in any way. I've pictured myself, in my head, as thin and svelte. I imagine what I would look like in certain outfits and clothes, and I have eaten certain foods, pushing what they'll do to my body to the back of my mind. I've imagined cutting off the fat parts of my body. I have denied that I'm treated any differently than the rest of the world, until it's slapped in my face by a disgusted look or seemingly innocuous comment.'[4]

Burdened by my new and utterly unfamiliar bulk it will be a struggle to do things I normally undertake without a second thought. Everyday actions like walking, climbing stairs, travelling on public transport or eating in public. Thinking about the day ahead, I wonder if I too risk being 'slapped in my face by a disgusted look'? Time will tell.

University of Sussex Campus: 10.20 a.m.

A passing student has shocked me by saying how 'disgustingly' fat I am to his companions, in a whisper I feel sure he wanted me to overhear. I caught this hurtful comment whilst struggling to retie laces on shoes I can no longer see – my bulging stomach blocks out the sight of everything below my waist. Until that moment, I had never realised the degree of precision and hand–eye coordination required by the everyday action of doing up one's shoes. Nor how tricky the task becomes when your feet are invisible.

I have only been obese for just over an hour and I'm already finding my new body hard to cope with. While, at 40 lb, this kapok-filled 'fat suit' is still far lighter than the same amount of fat tissue would be, it limits my movements and restricts mobility to a similar extent.

Heading for the Station: 10.45 a.m.

My next challenge is to negotiate a steep flight of concrete steps leading from our laboratory to the road. Walking down them requires a careful kind of faith. Everything feels so far away . . . my feet, my fingertips, the handrail. I decide not to attempt to grasp the handrail for support and instead focus my whole attention on the steps, suddenly terrified of losing my balance and taking a bruising tumble.

The phrase: 'Eat Less Move More' runs through my mind. How, when even walking is transformed into a daunting, exhausting task?

Arriving at the Station: 11.20 a.m.

It has taken me over half an hour to cover a distance I could normally walk in fifteen minutes. Passing a refreshment van outside the station, I notice two smartly dressed businessmen staring at me. Rather than return their gaze, which my confident, normal-weight self would almost certainly have done, I hastily look away. It's the same as I purchase a rail ticket. I avoid the eyes of the clerk, dreading that I might see contempt or pity in them. Negotiating the automatic barriers onto the platform is a struggle. I'm grateful no one else is following as I wriggle through.

On the Platform: 11.25 a.m.

I pass three women sitting on a bench and wonder whether they are looking in my direction, but avoid making eye contact. Later, studying the videotape, I see that they were staring with expressions of amused astonishment.

Ahead is a footbridge that has to be climbed, crossed and descended to reach the platform on the other side. If going down a flight of stairs immediately after leaving the lab filled me with dread, going up another proves even more of an ordeal. I have only climbed half a dozen steps before I have to pause to catch my breath. Twenty more to go. I feel as if I am climbing Everest. Less than three hours into my day, I already feel exhausted, embarrassed and discouraged.

On the Platform: 11.30 a.m.

Some of the passengers are glaring at me. So this is what it's like to look different! Another insight into what an ongoing battle it is to maintain a positive self-image when seriously overweight. I can see another overweight woman coming down the stairs. I give her a friendly smile but she looks right through me.

My train is just pulling into the station. As it slows, a teenage boy makes a rude face at me through the carriage window. I feel a surge of anger. How dare he! I know, though, that such hostility is encountered so frequently by those who are significantly overweight that they come to regard the judgemental looks, rude stares, and disparaging remarks as a normal part of life. Emma Burnell, a popular blogger who has struggled with being morbidly obese for years, writes about her own painful experiences. She speaks of exactly the kind of 'pointing, tutting . . . and judging' I am encountering now. For her, it happens 'daily, hourly – every single day.'[5]

Figure 1.1. Margaret waits for the train.

On the Train: 11.34 a.m.

The moment has come to board the train but I am reluctant to assert myself. I feel I must allow the other passengers to get on ahead of me. I am deeply reluctant to appear pushy, cause inconvenience to other passengers or draw attention to myself. Fortunately, at this time of the day the train is not very crowded. I mull over the fearful prospect of travelling during peak times. Someone would have to sit beside me and would probably feel either embarrassed or resentful about it, or so I pessimistically think. I make up my mind to avoid the rush hour on my return. I have found a seat to myself. I sway with the motion of the train and gaze out of the window.

At Brighton Station: 11.50 a.m.

We're just pulling in. I can't hop easily off the train and onto the platform, as I would do any other day. Instead I step off gingerly, holding onto the handrail. I waddle rather than walk down the platform. Approaching the ticket barrier, I start fumbling through my pockets to find the ticket. While I'm gradually getting used to an entirely different body, I can't believe how ridiculously uncomfortable it is to actually use my pockets; I soon decide they are out of the question and resolve to keep everything in the bag I have on my shoulder.

At this mainline station the barriers are busier than at the one where I boarded the train. There are passengers behind me who will be watching my struggles to squeeze through. Should I go through the larger barriers meant for people with luggage, or attempt to negotiate my vast behind through the regular ones? I decide that I will wait for the crowd to disperse before attempting to exit the station, and that I will use the regular barriers rather than walking across the platform to the roomier ones.

Passengers waiting on the concourse seem to be staring. I can't be sure because I'm afraid of looking at them directly, of catching their eye and seeing . . . what? Astonishment, shock, surprise? Perhaps I am becoming super-sensitive, but I am starting to get a sense of the constant compromises an obese person is forced to make.

Having to wait for the lights to go red when crossing the road, while others are able to cross whenever there is a gap in the traffic, confident that they can do so in good time; feeling compelled to hide away at the back of a café rather than subjecting themselves to the critical glances of strangers; having to reflect on every action rather than being able to make spur-of-the-moment decisions as a slim person can.

Window Shopping: 12.25 p.m.

*Figure 1.2. Margaret surveys the range of clothes in
a high street shop window.*

I'm slowly making my way through the town when I catch sight of my reflection in the glass of a shop window. I think how unattractive and drab I look. The range and prices of clothes available to average-sized and slender women is enviable. My outfit, which comprises a pair of trousers, a shirt and a jacket, cost me £125 in total. I could not find a jacket to fit my large body for less than £70, but here at Top Shop I can see coats for £25 and jeans at £30.

Not only are they good value but they look stylish and colourful. So far the items I have seen that I think might fit me have been mainly drab ponchos and tacky cover-ups – it is hard to imagine getting excited about wearing them or feeling attractive in them.

Resting on a Street Bench: 13.10 p.m.

Window shopping and strolling through the streets has been tiring but relatively non-stressful. People mostly left me to my own devices. Now, as I sit to catch my breath and gather my strength, the number of judgemental stares cast in my direction is unbelievable. Some passers-by stop and glare, others glance back to gawp at me, shaking their heads in disbelief. I can almost hear them saying to one another: 'Look at her, so fat and lazy she can't even be bothered to move.'

What really shocks me is the discovery that older people are far more likely to gawp and be judgemental than the young. I expected children and teenagers to be rude and cruel. I never anticipated such disrespectful behaviour from people my grandmother's age.

At an Open-Air Café: 13.45 p.m.

As a thin person I can get away with eating whatever I please without fearing the judgement of others. Today I play it safe with a Diet Coke and a packet of vanilla wafers. Buying these was not as difficult as I had feared; I did not receive any offensive looks when ordering. Now that I am squashed into a chair at a table as far from others as possible, some of my fellow diners are scarcely bothering to conceal their curiosity. I sit here sipping my Coke, feeling hungry but too intimidated to more than nibble tentatively on a wafer. I know I am being silly and that I should be able to rise above their unwelcome interest in me. Probably, given time, I would learn to ignore them. But at what cost to my confidence and self-esteem?

While conducting research into obesity I had often lamented the fact that many of my obese subjects were too inhibited to eat in the laboratory. Now I understand why. Even drinking my Coke subjects me to disapproving glances. I almost want to stuff a cupcake in my mouth just to spite them. As the café empties I continue to sit here, not wanting to go back onto the streets and subject myself to further unpleasant scrutiny.

Strolling through the Town: 15.15 p.m.

I finally screw up my nerve to start walking through the town again. I avoid looking at other pedestrians and keep my eyes fixed on the ground a few yards ahead. That way, if there are expressions of disgust, hostility or plain astonishment on their faces, I won't see them. Probably most people take no notice, just ignore me. But perhaps they won't. I decide not to risk it anyhow. I am becoming even more paranoid! How long would it take to get used to being obese, I wonder. A week? A month? A year? Never?

What sort of psychological barriers would I need to survive in such a hostile environment? Perhaps develop feelings of hostility to others. Maybe I'd make the first derogatory remark about my weight – laugh at myself before others could laugh at me. Or perhaps I would become hardened to what other people thought, ignoring them while silently cursing them for showing such a lack of respect to a fellow human being.

At the Skateboarding Park: 17.35 p.m.

My feelings of ostracism have really hit home. I'm walking past an awesome playground, play toys for younger children and a skateboard centre for adolescents. It's a place where, as my slim self, I can easily imagine spending a few hours on a sunny afternoon. Gazing at people roller-skating and jogging through the park, I am sad and irritated. It's so easy for those normal-weight people to suggest physical exercise as a way to lose weight. This may work for those only moderately fat. In my obese condition, prolonged activity of any kind would be out of the question. I realise that these emotions are, in a sense, ridiculous, since my current obese condition is not permanent. However, even the short time I have spent experiencing life in public in this state has been enough to make me feel genuinely hurt and frustrated.

As I pause to watch some roller-skaters, two teenagers openly jeer and shout at me. I am suddenly afraid. As a slightly built woman, I occasionally get pushed and shoved on crowded pavements. Now, being so much larger, I might have expected to feel safer. Yet confronted by these verbally hostile youngsters I am starting to panic. If I wanted or needed to run away I would not be able to do so. To be obese is to experience a vulnerability I've never previously known.

This experience is enough to make me head back to the station so that I can return to my normal size.

Back at the Lab: 19.35 p.m.

I have taken off 'the suit'. The experience of becoming obese, even for a day, was far tougher and more disturbing than I had anticipated. People are rightly outraged when minorities are bullied or abused, yet where 'fatism' – itself an absurd term – is concerned, apparently anything goes. It's always open season on the overweight. My heart goes out to those who, unlike me, cannot shed their weight in just a few minutes.

Had I been genuinely obese, I would doubtless have adjusted in time to the physical limitations imposed by my outsized body. My increase in weight would have been gradual, enabling me to come to terms with the restrictions it imposed over months and years rather than, as when donning a fat suit, in minutes.

'The disease of obesity, for it certainly is a disease, creeps on so slowly that the individual becomes thoroughly wedded to the form of life producing it before he realizes his true condition,' wrote America's first dietician, Sarah Tyson Rorer, in the late nineteenth century.[6] Over a century later, her words are as true today as they were then.

Yet, I know that if my weight did get out of control, I would feel terrible, having experienced a little of what it can be like to be overweight. If I was then unable to slim down again, those feelings would most likely change to depression, despair and, finally perhaps, weary acceptance. Maybe, like many of the overweight men and women I have met in the course of my work, and interviewed, I would have resigned myself to my fate and resorted to comfort eating to ease my psychological discomfort.

At this point we should make it clear that just because I, rather than David, my male co-author, became obese for the day, this does not mean we consider obesity an issue solely affecting women. Far from it – we would argue that both men and women have an equal struggle to maintain a healthy body image, whether obese or lean. At the same time, obesity is a gender issue, and it is a class issue. While men of upper and working classes are equally overweight, it is astonishing to see how class and professional status impact on women's weight. For example, for women living in the UK, the prevalence of obesity doubles depending on professional status: 35% of women in unskilled trades are obese, versus 18% of those on a professional path.[7] One key difference here seems to be that women see themselves – and tend to be seen and judged by others – somewhat differently to men. This is something we will discuss further in Chapter 3.

The judgemental looks that were such a constant during my day out in Brighton reflected a view widely held in society: that obesity is due to the greed and laziness of the obese person. However, the truth is that putting on weight has less to do with individual weakness than to the fact that we are living in an obeso-genic environment. That is to say, one in which multiple forces, many outside our personal control, encourage the consumption of foods that are calorie rich but nutritionally impoverished. In the next chapter we will explain what some of these forces are and discover why people who struggle with maintaining a healthy body weight are not to blame for their condition.

CHAPTER 2

'Hey fatty, get off this train!'

'If you think it's simple, then you have misunderstood the problem.'
Bjarne Stroustrup, Texas A&M University[1]

'You big fat pig,' were the words twenty-two-stone Marsha Coupe heard before being violently assaulted. Not by some drunken thug but by a respectably dressed middle-aged woman on an evening commuter train.

'I was returning home . . . and a woman sitting across from me started kicking me,' Marsha recalls. 'She said, "Hey fatty! You should not be on the train, you need two seats!"'[2]

Fifty-three-year-old Marsha, who suffered severe bruising and who doctors feared might lose an eye, was shocked but not especially surprised by the unprovoked attack: 'Fat people are fair game for everyone,' she explained wearily. 'I've had beer cans thrown at me by youngsters, but the abuse doesn't just come from the obvious places. The normal rules about behaviour, respect and common courtesy don't apply to us.'[3]

Her attacker fled before the police could be called.

'There is true aggression towards overweight people and it comes down to fear and a complete lack of understanding of the issue,' says psychologist Ros Taylor. 'People think, "I can control what I put in my mouth so why can't they?" But we're not all the same, we don't all start from the same point.'[4]

According to the Americans with Disabilities Act of 1990, anyone who is significantly overweight can expect to 'continually encounter various forms of discrimination, including outright intentional exclusions, the discriminatory effects of architectural, transportation, and communication barriers, overprotective rules and policies, failures to make modifications to existing facilities and practices, exclusionary qualification standards and criteria, segregation, and relegation to lesser services, programs, activities, benefits, jobs, or other opportunities.'[5]

'Society's increasing hatred of fat and obsession with thin is creating appalling prejudice,' agrees psychologist Susie Orbach, author of *Fat Is a Feminist Issue.* 'It is allowing people to feel justified about abusing fat people.'[6]

Many people with a weight problem accuse politicians and the media of encouraging intolerance by promoting obesity scare stories that provoke disgust, mockery and moral outrage.

'The government and the press have created an atmosphere where people think they have a legitimate right to go up to an overweight person and tell them how to live their lives,' says Marsha Coupe. 'To them we are all the anonymous pictures of fat people they see in the papers and are the cause of all society's ills, as well as a drain on the NHS. We deserve what we get. We're not people with feelings.'[7]

Research by Rebecca Puhl and Chelsea Heuer, from the Rudd Center for Food Policy and Obesity at Yale University, strongly supports this view. 'Numerous studies have documented harmful weight-based stereotypes that overweight and obese individuals are lazy, weak-willed, unsuccessful, unintelligent, lack self-discipline, have poor willpower, and are noncompliant with weight-loss treatment,' they report. 'These stereotypes give way to stigma, prejudice, and discrimination against obese persons . . . Perhaps because weight stigma remains a socially acceptable form of bias, negative attitudes and stereotypes toward obese persons have been frequently reported by employers, co-workers, teachers, physicians, nurses, medical students, dieticians, psychologists, peers, friends, family members, and even among children aged as young as three years.'[8]

'Day after day, you're terrified'

Rebecca Rees and her colleagues from the Institute of Education at the University of London have revealed that being young and overweight can transform growing up into a nightmare. After analysing thirty studies involving young people aged between twelve and eighteen, and going back almost twenty years, they uncovered a terrifying litany of verbal and physical bullying on a daily basis. This included 'being threatened with a knife, beaten, kicked, pushed down stairs and having objects thrown at them.' More commonly, overweight children had to endure verbal and social abuse such as name-calling, deliberate and extended isolation, whispering and sniggering. Much of this occurred in school and, in some instances, left them so scared they refused to attend.[9]

Physical Education lessons were cited as a particular source of hurtful and humiliating ridicule. 'Day after day, you're terrified,' one youngster told researchers miserably.[10]

Unsurprisingly, these attacks impacted negatively on the emotional health of the victims, reducing confidence and increasing loneliness, depression and anxiety, especially when it came to visiting public spaces. Shopping trips and other social events often left them feeling marked out as different and excluded. 'I . . . just wanted to be part of the crowd and not to stick out like a sore thumb,' Huw, an overweight seventeen-year-old reported. 'Because sticking out . . . when someone sees the person who looks, who is bigger than . . . almost everyone there, that makes you feel really bad.'

The researchers found that being large did not make youngsters any more tolerant of those who were similarly overweight, however.

'Fat people, I hate fat people,' one teenage Scottish girl told them.
'I don't hate their personalities, I just don't like the way they look.
I just don't know why folk would do that to themselves.'[11]

Generally, overweight young people saw others with large bodies
as being either lazy or incapable of controlling their greed – or
both. For all of them the ideal body size for a young woman was
thin, while for a man it was muscular and fit-looking. As with
adults, any failure to match these ideals tended to be blamed on
a lack of self-control. In one study 'participants stated "quite
fervently" that a young person's size was their own responsibility,'
comments Rebecca Rees, 'and in only two studies did young people
suggest that some responsibility might lie elsewhere. Young people
who felt, or already were large made it clear that they knew they
had to do something and tended to be critical of their own self-
will.'

The researchers concluded, sadly, that: 'The perspectives of
young people in the UK . . . paint a picture of a stigmatising and
abusive social world.'[12]

Coming to Terms with Weight Gain

Douglas Degher and Gerald Hughes, from Northern Arizona
University, report that the development and acceptance of a 'fat
identity' occurs in several well-defined steps. It often starts with
teasing and bullying at school, leading to the victim identifying
themselves as a 'fat boy' or 'fat girl'.

This can then lead to a range of psychological coping mechanisms, such as denial ('I am not fat I am just a bit chubby') and
avoidance ('I refuse to think about my weight') to compliance ('I
will go on a diet not for myself but to please others').[13]

And, as Dr Ian Campbell – a specialist at the Overweight Clinic
at University Hospital, Nottingham, and honorary medical director

of the charity Weight Concern – points out, 'The result is the people who need the most help don't seek it. They are left feeling guilty and undeserving.'[14]

While (as we explain in Chapter 3) there are people who take pride in being seriously overweight, research and practical experience suggest that most do not. When Dr Colleen Rand, a cosmetic surgeon at the University of Florida, asked forty-seven formerly obese men and women whether they would sooner have some disability rather than regain their previous weight, the replies were surprising and shocking. All said they would rather be deaf than obese, nine out of ten (91%) preferred to lose a leg or even go blind (89%)! As one put it: 'When you're blind, people want to help you. No one wants to help you when you're fat.'[15]

Small wonder then that anyone with a serious weight problem can become discouraged, depressed and even despairing. It is a cruel irony that these are emotions that can lead to comfort eating, as we explain later, in Chapter 9.

But before we go on to look at the real reasons why the world is becoming obese, we should spend a few moments considering what precisely this term actually means. What, for example, is the difference between being overweight and obese? How is the distinction made and, even more significantly, how accurate is the main method used to determine it?

Is Arnold Schwarzenegger Obese?

At 6 feet 2 inches tall and weighing, at his heaviest, 260 lb, the Austrian/American actor and one-time governor of California has a Body Mass Index (BMI) of 33.3. The clinical definition of a lean or rather 'normal' BMI is anything between 20–25 kg/m². The range 25–29.9 kg/m² is considered overweight, and the cut-off for obesity is 30 kg/m². This would seem to mean that a man once described by the *Guinness Book of World Records* as 'the most perfectly developed in the history of the world' was severely obese!

Which is, of course, nonsense.

The error arises because Schwarzenegger, who began his career as a weightlifter and still works out daily in the gym, is so muscular. When it comes to calculating BMI, it is the weight of his skeletal muscle rather than adipose tissue (muscle is about 18% denser than fat), which results in such an absurd conclusion.[16] It also highlights one of the main problems with using BMI as a measure of any individual's fatness. So what exactly is the BMI, why was it invented and how is it currently being used and abused?

Why the BMI is a Great Big Lie

The statistical technique for comparing people's weight was invented by the eighteenth-century Belgian statistician and polymath Adolphe Quetelet.[17] At the request of the Belgian government, who wanted a quick and easy technique for determining obesity levels among the general population, he created a standardised measure based on an individual's weight divided by the square of their height. It was an assessment technique that, he always emphasised, could, and indeed should, only be used to assess groups and never individuals.

In 1972 American physiologist Ancel Benjamin Keys renamed this measure the Body Mass Index or BMI.[18] Today the formula for obtaining BMI (weight in kilograms divided by height in metres squared) is widely used by doctors and scientists to place people into one of five weight classifications – in spite of the fact the measure was never intended to be used on an individual basis in this way.

Anyone with a BMI of 18.5 or less is considered to be underweight.

A BMI of between 18.5 and 24.99 indicates normal weight.

A BMI of 25 to 29.99 means the subject is overweight. A BMI from 30 to 34.99 indicates severe obesity and, from 35 to 39.99, morbid obesity. Anyone with a BMI of 40 or higher is classified as super obese.[19]

The above categories apply to Caucasians only, with other parts of the world adopting different cut-off points. In Japan, for example, anyone with a BMI greater than 24 is deemed to be overweight,[20] while in China being overweight is indicated by a BMI greater than 23.[21]

While perfectly valid when applied to groups of people, when used to determine an individual's weight the BMI becomes, in the words of Stanford University mathematician Professor Keith Devlin, 'mathematical snake oil'. A perfectly healthy, athletic person can easily be categorised as obese in this system because they have more heavy muscular tissue. In spite of this drawback, BMI is widely, if unwisely, trusted. Devlin comments that: 'Because the BMI is a single number between 1 and 100 (like a percentage) that comes from a mathematical formula, it carries an air of scientific authority.'[22]

As we saw with the case of Arnold Schwarzenegger, it fails to take account of the relative proportions of bone, muscle and fat, so the results may defy logic. Since bone is denser than muscle and twice as dense as fat, a combination of low fat with strong bones and good muscle tone will still produce a high BMI.

So while the formula works reasonably well for sedentary people, whose bodies combine a high relative fat content with low muscle mass, it gives entirely the wrong answer in the case of those who are fit, healthy and lean. It also very misleadingly suggests that there are precisely five main weight categories (under, over, obese, morbidly obese and just right) whose precise boundaries depend on a decimal place. Finally, BMI cannot account for allocation of adipose (fat) tissue. As scientists discover more about the function and activity of fat, they have become adamant that the proximity and quantity of fat to organs bears more relevance to health than a person's body weight.

So the bottom line is that no one should take any notice of their BMI when assessing how fat they are. Remember: it was never intended to apply to individuals, just to groups of people. There are, admittedly sometimes more complex and costly, ways of determining body fat with great accuracy. They include skin-fold measurements, underwater weighing, bioelectric impedance, functional Magnetic Resonance Imaging (fMRI) and infrared detection.

A simpler rule of thumb is to use your waist circumference. For men, if this exceeds 40 inches, there is a high probability of obesity; for women the measurement is 35 inches or above.[23]

Who's Really to Blame for Obesity?

Ask someone why a person is overweight and they will most likely reply: 'Because they eat too much and exercise too little!'

However, this is a crass oversimplification of what is likely to be a much more complex issue. When someone has a serious weight problem it is not an issue of *culpability* but of *vulnerability*. They have become, for reasons we will explain later in the book, hypersensitive to ubiquitous environmental food cues and the addictive rewards provided by foods rich in sugar, fat and salt.

While a combination of poor eating habits and insufficient exercise are undeniably part of the problem, they are by no means the whole or even the main reason for the obesity pandemic.

Ultimately, it has arisen because our modern lifestyles are dramatically out of step with our biology. We are all equipped with a digestive system that never evolved to cope with the easy and unlimited availability of high energy-dense (HED) foods we see today. Millions now work in stressful conditions, in largely sedentary occupations, under time pressures that enforce rushed meals augmented by energy-dense snacks. Many people are too time-poor and tired at the end of the day to engage in strenuous exercise, and many get insufficient sleep. We live in an obesogenic environment characterised by relentless and inescapable advertising and marketing of food.

In short, obesity primarily results not from self-indulgence but from a mismatch between natural physiological needs and conditioned psychological wants.

CHAPTER 3

Fashions in Fat

'One can never be too thin or too rich.'
Wallis Simpson, Duchess of Windsor[1]

On Thursday 23 June 1809, a most unusual funeral took place in the small Lincolnshire town of Stamford, some ninety miles north of London. The elm coffin was six feet long, four feet wide and fitted with wheels. Daniel Lambert, the 39-year-old who lay within, weighed 52 stone and was England's fattest man. In order to remove Daniel's body from his lodgings, a window and wall of the building had to be demolished. So heavy was the load that, at the cemetery, a gently sloping trench had been excavated to enable the sweating pall-bearers to push the massive coffin into the grave, rather than attempting to lower it. Even so, it took twenty strong men ninety minutes to manoeuvre Daniel into his final resting place.

What is interesting about the story of Britain's fattest man is not so much how he managed to become so obese as the way he was perceived by others.

As a youth, Daniel had been tall and stocky but by no means overweight. He was an active young man who could walk for miles without effort and who enjoyed lengthy swims. At the age of eighteen he had, like his father before him, worked as a jailer at the local prison. It was from this time that his weight rapidly increased, although quite why it did so remains a mystery. The doctors who examined him could find no medical reasons for his prodigious girth and Daniel himself always insisted, and the contemporary literature confirms, that he drank and ate modestly, never having 'more than one dish at a meal'.[2]

Surprisingly, given his size, he appears to have remained fit. He was reported as being in 'perfect health: his breathing was free, his sleep undisturbed, and all of the functions of his body in excellent order'. After his death, he was remembered as 'a man of great temperance.'[3]

Whatever the cause of his obesity, by the time the jail was closed and he lost his job, Daniel was too fat to find further work in his home town. In the hope of finding employment further afield, he travelled to London, where he soon fell on hard times. Almost destitute, he was obliged to exhibit himself in public.

It cost a shilling (5p) to visit him at his London home, to ask him questions and stare in astonishment at his prodigious size and marvel at the size of his clothes – six grown men were able to fit inside his waistcoat at the same time.

Intelligent, knowledgeable and a witty conversationalist, Daniel soon became a fashionable attraction. One banker was so intrigued he made twenty visits, while a party of fourteen made the trip all the way from Guernsey. He was even invited to an audience with King George III.

After a few years, he had become sufficiently wealthy to return to Leicestershire, spending his few remaining years breeding and selling hunting dogs. He died suddenly, on 21 June, at his lodgings in the Wagon & Horses on Stamford High Street. Friends and admirers raised the money to pay for a funeral and provide a headstone on which they had carved: *Remembrance of that prodigy in nature Daniel Lambert, a native of Leicestershire who was possessed of an exalted and convivial mind and in personal greatness had no competitor: He measured three feet one inch round the leg and weighed fifty two stone eleven pounds. He departed this life on the 21st June, 1809, Aged 39 Years.'*

As this memorial testifies, Daniel was regarded by his contemporaries not as a glutton to be condemned but as a 'prodigy of nature' to be admired. During the early decades of the nineteenth century, as for several centuries before, being extremely fat was not merely acceptable – it was fashionable. Being overweight symbolised power, dominance, wealth and even hypersexuality; it could be seen as a source of pride, not a cause for dismay.

Fat and Proud

'Obesity', observes historian Peter Stearns of George Mason University, 'was . . . associated with sound health in a period when many of the most debilitating diseases, such as tuberculosis, led to the body wasting away.'[4] In Britain during the last decade of the eighteenth century, the fashion for gluttony and obesity was led from the top by the Prince Regent, later to become George IV. By his forties a combination of heavy drinking, vast banquets and lack of exercise had taken his weight to more than 17 stone (245 lb) and his waist to 50 inches (130 cm). Edward VII, the son of Queen Victoria, who followed his mother to the throne after her death in 1901, was almost equally stout, so much so that he was apparently unable to fasten the bottom button of his waistcoat. The story goes that to make their monarch feel better about this, his court followed suit, setting a fashion which endures to this day.

Where women were concerned, being plump and voluptuous, with milk-white skin – 'Rubenesque' – demonstrated not only that you were wealthy enough to eat in excess but also that you were able to lead a life of ease.

Nor did 'fat pride' completely disappear as improved public health and medical science began to banish wasting ailments. In 1903, a group of American travelling salesmen established the Fat Men's Club of Connecticut, such was their pride in being overweight. Membership was only open to men weighing at least 200 lb, and members competed to see who could put on the most weight from one year to the next. The club, whose slogan was 'We're fat and we're making the most of it!' survived until 1925 – several years longer than most of its founding members![5]

When Dieting Became the Vogue

However, concerns about the consequences of being seriously overweight did begin to surface during the late nineteenth century. In 1863 William Banting published one of the Western world's earliest diet books, *Letter on Corpulence Addressed to the Public*. It was a first-hand account of how he, a successful, middle-class undertaker (his firm made the coffin for the funeral of the Duke of Wellington) had lost 35 lb by going on a diet. Such was the popularity of his book, in which he reported, 'I can honestly assert that I feel restored in health, bodily and mentally' that the word 'banting' quickly became synonymous with dieting in much of Europe.[6] In Swedish *bantna* still means 'to diet'.

Of course, Banting's book was not the first to advocate dieting; the word diet comes from the Greek *diatta*, meaning the 'manner of living', and discussions of the importance of good diet first appear in the writings of the early physicians such as Hippocrates and Galen. And the world's oldest cookery book, the ten-volume Roman *Apicius*, dating from the fourth or fifth century AD, contains much dietary advice which still applies today.

One volume, devoted to pulses (legumes), comments on the benefits of a vegetarian diet, as expressed in *The Book of Daniel* (1:8–16). We learn how Daniel and other children from Judaea, after being brought to the palace of Nebuchadnezzar, King of Babylon, were offered meat and wine. They asked, instead, to be allowed to eat their traditional meal of pulses. Their guard, the prince of the eunuchs, feared that by doing so the youngsters would look less well-nourished than other children eating meat and drinking wine, and that if this proved to be the case he might lose his head. Daniel suggested an experiment. He and his companions would consume pulses and drink water for ten days, after which the king should compare their apparent state with that of 'the children that eat the king's meat.' This was agreed and, at the end of ten days, 'their countenances appeared fairer and fatter in flesh than all the children which did eat a portion of the king's meat' (Daniel 1:15).

Centuries later we hear of a very famous man attempting another kind of diet. In 1067, the year after he had invaded England and defeated King Harold and his army outside Hastings, William the Conqueror became one of the first notable names in history to try to lose weight by going on a liquid diet. He did so by retiring to bed and drinking only alcohol!

'With the publication of *De re Medicina* in 1478 in Florence, Italy, diet became an important part of medical practice,' write June Payne-Palacio and Deborah Canter in their book *The Profession of Dietetics*. 'Medicine was divided into three branches – diseases treated manually, diseases treated by medicine, and diseases treated by diet. In 1480, the first printed cookbook appeared, containing reference to quality and variety of meat, fish, fruits, and vegetables; information on how they nourish the body; and directions on how they should be prepared.'[7]

So, Banting's book was really part of a long discourse on dieting that had been going on for centuries. However, despite the rapid growth of books on dieting following this bestseller, and despite increasing scientific interest in the topic of obesity, being fat remained a desirable trait amongst an exclusive circle of the powerful and influential; an indication that the individual could be trusted and relied upon.

There are still many countries in which being overweight

remains fashionable because it is associated with social status and good health. In South Africa, for example, where nearly 40% of men and 70% of women are overweight or obese, a recent survey found that almost nine out of ten people (88%) believe 'fat is beautiful'. The plumper a woman, the more sexually attractive and healthy she is considered to be. A man with a big belly (in Afrikaans a *boep*) is similarly viewed as healthier, wealthier and more successful than a slender male, whose thinner frame is regarded as a sign of poor health and a lack of financial success.[8]

But in much of the world, the attitude towards heavier people is far from being this positive.

When Fat Fell Out of Fashion

Although the word 'obesity' first appeared in a French-English dictionary in 1611, attitudes towards being fat only started to change significantly during the early years of the Roaring Twenties. This was the decade that saw the emergence, in the West, of a 'new breed' of young women. Known as 'flappers', they wore short skirts, smoked and drank heavily, drove cars, listened to jazz and enjoyed casual sex. They were active and favoured an androgynously slender physique.

'One can never be too thin or too rich,' pronounced Wallis Simpson, Duchess of Windsor, and in Britain, at least when it came to slimness, women strove to obey her diktat.

Since then, and despite the obesity pandemic, being fat has increasingly fallen out of favour across the developed world.

Obesity was soon regarded by doctors as a threat to the nation's health, and by moralists as evidence of the nation's decadence. Gluttony was one of the Bible's seven deadly sins; nature's punishment came in the form of bulging bellies, dimpled thighs and flabby buttocks.

This new puritanism spread like wildfire through Western populations, with people becoming ever more judgemental of the overweight. It is an attitude which remains deeply ingrained to this day; when, in 2004, Sander Gilman from the University of Illinois showed children silhouettes of fat and slender people, she found they were more likely to perceive the former as 'lazy', 'dirty', 'stupid', and as people who 'cheat' and 'lie'. According to Gilman's research: 'Physicians are not much better than the children. They describe their obese patients as "weak-willed, ugly and awkward."'[9]

This rather dismal view is supported by the findings of a study by Melanie Jay and her colleagues from the New York University School of Medicine. They report four out of ten doctors feel less motivated to help significantly overweight patients whom, they believe, are less likely than slender ones to benefit from their advice.[10]

Many primary care physicians rated therapies aimed at altering patient behaviour as less effective than treatments such as pharmacological intervention and surgery for nine out of ten chronic conditions relating to obesity, with less than 50% feeling confident in prescribing weight-loss programmes, and fewer than one in six (14%) describing themselves as 'usually successful' in helping obese patients lose weight.[11] Even health professionals specialising in obesity had an 'implicit anti-fat bias and are more likely to automatically associate "fat people" with negative stereotypes than "thin people"'.[12]

In another study, three out of ten (31%) of internal medicine residents reported that treating obesity was futile, with only four out of ten (44%) considering themselves qualified to treat such patients.[13]

'There is strong evidence,' comments Melanie Jay, 'that physicians do not counsel obese patients adequately and attitudes may be one reason for this deficiency.'[14]

Fat as a Dieting Issue

We live in an age of dietary extremes. While waistlines are expanding fast, all around the world, so too is the number of people obsessed with becoming or remaining thin. Every year, in the UK and US alone, millions of people invest vast amounts of time and money, endure discomfort and suffer pain in the hope of retaining or regaining a slender figure. They go on diets, undergo cosmetic surgery, join gyms and health clubs, send their overweight youngsters to fat camps, pop pills, swallow supplements, consume special foods, smear on lotions, snap up the latest diet books, watch dieting DVDs, browse weight-loss magazines, seek out personal trainers, consult dieticians and join slimming clubs.

In Britain the diet industry is worth £2 billion, only slightly less than the £2.33 billion spent by the National Health Service on its nationwide A&E services.[15] In the US, the diet industry is worth in excess of $60 billion.[16] In 2009, over 400,000 Americans went under the knife in their quest to, literally, lose weight. This involved 354,015 liposuctions and 713,115 operations for abdominoplasty (tummy tucks). In the UK the figures are more modest: in 2011, 3,581 people underwent liposuction and 3,375 tummy tucks.[17]

Paul Campos, a law professor at the University of Colorado, has suggested that obesity has become a device to promote intolerance, which shows 'disturbing similarities to the eugenics movement, with its emphasis on "improving" the species'. He characterises the 'war on fat', as a pivotal time in American history, representing 'the first concerted attempt to transform the vast majority of the nation's citizens into social pariahs, to be pitied and scorned.'[18]

'The things many Americans worship today – "health", "fitness", a perpetually youthful body,' Campos writes in *The Obesity Myth*, 'have become so closely associated with staying or becoming thin that, for all practical purposes, what such people worship is a god of perpetual slenderness.'[19]

Yet, as we explain in Chapter 13, losing weight, and making sure it stays lost, is far from the straightforward task that producers of diet plans and peddlers of slimming pills try to make out. Up to half of the weight shed has often been regained within a year, and in time the vast majority of those whose diets have succeeded in the short term will either have returned to their previous weight or actually become even heavier.[20] Over the past thirty-three years, there has not been a single report of a country successfully reducing obesity rates among its population.[21]

Fat as a Feminist Issue

For many critiques of the 'war against weight', the emphasis on female slenderness is far from accidental. They argue that the early 1960s, when the dieting movement really took hold (Weight Watchers, for example, was founded in 1961 by an overweight New York housewife named Jean Nidetch), marked the start of the sexual revolution.

For the first time, the idea that a purely domestic role – 'die Kirche, Küche, Kinder' (church, kitchen, children), as the Germans say – should be the only one a woman should crave, began to be seriously and widely challenged. Women felt able to openly express and explore their appetite for sex.

At the same time, as we have already noted, the emphasis on the importance of physical appearance resulted in fatness being seen as less a health and more a moral issue. It was increasingly stigmatised as a sign of greed, laziness and lack of self-control.

'If you fail to lose weight you are demonstrating you're a bad person,' says Peter Stearns. 'It's a big burden. Faced with this additional pressure you are even more likely to end up by saying:

"The hell with it! I'm going to get ice-cream. I am such a bad person I need to solace myself."[22]

In 1978, psychotherapist Susie Orbach struck back against this bullying and mockery with her groundbreaking anti-diet book, *Fat is a Feminist Issue*. She argued that, by becoming obese, a woman was deliberately setting out to challenge gender stereotypes. Thirty years on, the same views continue to be espoused by feminists such as singer Beth Ditto. At 5 feet tall and weighing 15 stone, Beth vocally and publicly rejoices in her weight, challenging culturally prescribed norms and popular notions of what it means to be beautiful. Described by writer Germaine Greer as 'the coolest woman on the planet', Beth describes herself as a 'fat, feminist lesbian from Arkansas'.[23] She has posed nude on the front covers of *NME* and *LOVE* magazine, launched her own plus-size collection of fashions, and uses her obesity to promote her view that 'every person is beautiful in their own way'.

In the face of the mounting pressure for women to comply with ridiculous standards of beauty, 'fat acceptance' movements such as Fat Liberation, Fat Activism, and Fat Power have garnered widespread support, especially within the US. Programmes such as Healthy At Every Size (HAES) offer support for all those who want to challenge the current status quo. HAES 'acknowledges that good health can best be realized independent from considerations of size. It supports people – of all sizes – in addressing health directly by adopting healthy behaviors.' In theory, this should be the health message promoted by all.

Unfortunately, the credibility of many of these movements is undermined by research findings. For example, some groups claim that the risks of weight cycling (where weight is repeatedly shed and then gained once more) are significantly greater than those of being overweight. In fact the exact opposite is true. While weight cycling is certainly by no means ideal, the idea that it contributes to either hypertension or Type II diabetes, as some of these groups claim, has been disproved by studies going back well over a decade.[24] Another claim promoted by some fat advocacy groups is that such yo-yo dieting increases the risk of mortality. This too has been exposed as a myth by epidemiological studies, such as one conducted between 1992 and 2008, which involved more than 55,000 men and 66,000 women.[25] While it is true that being over-weight (but not obese) offers some health benefits to those over 65, once this age qualifier is removed, the purported 'advantages' of being excessively overweight disappear.[26]

Having said all this, it is certainly true that adipose tissue plays a vital role in maintaining good health, and this is something we will examine in greater depth in Chapter 5. But when present in excess it poses grave risks to health. As weight increases, so too does the likelihood of developing Type II diabetes, cardiovascular disease, and a series of cancers. Body dysmorphia (a preoccupation with perceived defects in one's physical appearance) and the mere idea of Fat Activism reveals a deep misunderstanding about the true nature of 'health'.

Promoting the supposed benefits of being either too thin or too fat is actively misleading and potentially dangerous. Only by coming to a clear understanding of the risks posed by excess adiposity can we begin to appreciate the very real threats to our health presented by the obesogenic environment in which we live.

In Part Two, we shall explore the biology of obesity to explain some of the reasons why people are becoming so overweight. We start, where life begins, with the newborn. For if being fat is currently out of style with adults, it is still very much in fashion with babies.

PART TWO

Inside Story:
The Biology of Obesity

'Biology is now bigger than physics, as measured by the size of budgets, by the size of the workforce, or by the output of major discoveries; and biology is likely to remain the biggest part of science through the twenty-first century.'
Freeman Dyson[1]

CHAPTER 4

Born to Become Obese?

'There is growing evidence that a baby's development
before birth has a major impact on their health in later life.
This means that the prevention of obesity needs to begin
in the womb.'[1]
Neena Modi

Human babies are the fattest animals on earth.

At birth between 15% and 16% of a baby's body consists of fat;
they continue piling on the pounds for the next twelve months,
consuming some 220,000 calories in the process.[2] By the time they
blow out the candle on their first birthday cake, up to 30% of
their body weight is made up of fat.[3]

Nor is this our species' only peculiarity where fat is concerned.
While other mammals, including our primate relatives, only
produce white fat once they have started to suckle, human babies
start doing so while still in the womb.[4] This explains why, to the
delight of their proud parents, they look so plump and feel so
smooth to the touch. By contrast, other newly born primates
have only 3% fat, and so look and feel, in the words of German
anthropologist Adolph Schultz, 'decidedly "skinny" and horribly
wrinkled.'[5]

But why are human babies so much fatter than other primates?
What function can this vast calorific investment in producing
adipose tissue possibly serve? And what can fat infants tell us about
the causes of adult obesity?

Fat for Sleeping and Fat for Swimming

Mammals require large deposits of fat for two main reasons – to hibernate or as protection from the cold. Hibernating animals build up their fatty reserves when food is plentiful and live off it when food is scarce.

In animals exposed to constant cold, such as those living in the oceans or in frozen wastes, fat is present all year round. And although only a few humans are ever exposed to such extreme environments, we too maintain large stores of fat from one season to the next. While most animals deposit fat deep within the body – especially between the muscles and around the heart, kidneys and intestines – both humans and sea-dwelling mammals store a high proportion directly beneath the skin. If you were to remove the skin of a rabbit or an ape, much of the superficial fat would remain behind on the organs. Do the same with a human, a seal or a dolphin and a large proportion would come away with the skin. The differences between *visceral* (deposited around the organs) and *subcutaneous* (deposited beneath the skin) adipose tissue are extremely important and we will return to them later in this chapter.

A number of theories have been advanced as to why we humans are so very different from every other land animal in the way we synthesise and deposit fat. We must assume that there is a good reason for this; if storing fat in the way we do was simply detrimental to our health, evolution would surely have corrected this by now. In this chapter we will explore several possible theories for the distribution of our fat tissue, as well as considering our brain-to-body ratio (or 'encephalisation quotient'); another key feature of human biology which makes us so unique among mammals. As we will see, the relationship between the brain and our evolved patterns of fat storage has critical relevance to the problem of obesity.

The Naked Ape Theory

The naked ape theory proposes that fat provides insulation for human infants which, unlike every other primate, are born both hairless and chubby.

'Fat [is plentiful] in the new-born baby,' wrote David Sinclair, Professor of Human Anatomy, University of Aberdeen, 'probably as an insulation against cooling.'[6] At first sight this seems to make a lot of sense. Human infants have a large surface area per kilogram of body weight.[7] A baby with an average birth weight of between 7 and 8 pounds, for example, will have a Body Surface Area (BSA) of around 5 square feet, about the same size as a picnic rug, which means they lose heat rapidly.

But while the 'insulation' theory sounds plausible, there is a problem with it. Although human babies have nearly seven times as much fat beneath their skin as elephant seals, and three times more than fur seals, they don't *really* need it to stay warm.

Seal pups, after all, spend their infancy swimming in freezing oceans or sleeping on sheets of ice in sub-zero conditions, while most human infants are snugly wrapped against even the slightest chill. And the Inuit, who inhabit some of the earth's most intensely cold regions, do not possess extra layers of fat beneath their skin to provide greater insulation. Even as babies they, and other people indigenous to the polar regions, have subcutaneous fat deposits either equal to, or even smaller than, those living closer to the tropics.[8]

This has led many to conclude it is the human's ability to generate heat via their lean body mass, rather than to conserve it by means of adipose tissue, that enables infants, as well as adults, to maintain a steady core temperature of 37°C.

'Humans are among the fattest of mammals,' observes anthropologist Professor William Stini of the University of Arizona, 'but do not adapt to cold climates to any significant degree through the acquisition of thick layers of subcutaneous fat.'[9]

So if layers of fat beneath the skin are not, as the naked ape theory proposes, to protect us from the cold, what are they there for?

The Aquatic Ape Theory

In the early 1940s, Max Westenhöfer, a medical doctor from the University of Berlin, described what he termed 'aquatile Mensch'.[10] Influenced by the mood of National Socialism of the time, Westenhöfer disagreed with aspects of the Darwinian theory of kinship between ape and man. Instead, he looked to the sea to describe traits unique to man, disputing the idea that we evolved strictly from plains-based primates. However, his theory was riddled with inconsistencies, and did not reach a large audience.

Some twenty years later a marine biologist, named Alister Hardy proposed a similar theory. 'The suggestion I am about to make may at first seem far-fetched,' he told an audience of SCUBA diving enthusiasts in Brighton, 'yet I think it may best explain the striking physical differences that separate Man's immediate ancestors (the Hominidae) from the more ape-like forms (Pongidae) which have each diverged from a common stock of more primitive apelike creatures which had clearly developed for a time as tree-living forms.'[11]

Hardy went on to expound his own version of what became known as the Aquatic Ape Theory. He argued that around 5 million years ago, due to competition from other animals our ancestors were forced to leave the plains to search for food in the oceans. He claimed that the length of our legs and our lack of hair are signs of us having adapted to an aquatic environment.

He also pointed to similarities between the adipose tissue deposits in humans and ocean-dwelling mammals, such as dolphins.

Sweating It

While superficially persuasive, the Aquatic Ape Theory is rejected by the majority of scientists today. A more popular explanation is that rather than helping to streamline the body for swimming, our lack of hair can be explained as the result of our need to control body temperature within fairly tight limits.[12]

Sweating is a relatively wasteful mechanism for temperature control, in that it uses up valuable water. In a hot environment such as a savannah, a furry animal would require even more water for cooling, since the fur would naturally cause it to retain heat and therefore perspire more. It seems likely that a perspiring animal that evolved in a hot climate would therefore gradually lose its fur. However, the loss of fur would likely lead to some kind of compensatory tissue, as the animal would still have to contend with cold temperatures upon occasion. This is where other scientists postulate our hairless appearance and our concentrations of subcutaneous fat unique in land mammals, have their origins – they allow us to stay reasonably cool when it is warm, and reasonably warm when it is cool.

Subcutaneous fat as a method for insulating an organism (rather than external fur) makes a lot of sense from an evolutionary perspective and we certainly should not discount the theory. However, there is also a strong possibility that it is our brains rather than our bodies that require the extra glucose which stored fat represents. When we consider our physiology in more detail, particularly in comparison to other mammals, there is good reason to think this may be the case.

Building a Bigger Brain

The average human brain is 1,000 grams larger than would be expected for a mammal of our size and weight.[13] At 20 watts, the metabolic cost of our brain is also more than four times greater than that of the typical mammal, which is around 3 watts.

'A human child under the age of 5 years uses 40% to 85% of resting metabolism to maintain his/her brain,' according to William Leonard and Marcia Robertson of the University of Guelph. 'Therefore, the consequences of even a small caloric debt in a child are enormous given the ratio of energy distribution between brain and body. Hence, the prolonged period of growth in humans may be partly an adaptation to limit the already high total and brain energy requirements during childhood.'[14]

Although these demands decline as the child develops (the adult brain consumes about 20% of total energy production), this still represents a considerable investment. The brain makes up just 2% of our body weight yet it consumes more than twenty-two times the amount of energy required by our muscles.[15]

Which raises the question of how humans are able to afford the metabolic cost of possessing such a large brain.[16]

Where does the energy required to fuel the brain come from?[17]

In 1995, Leslie Aiello, a Reader in Psychological Anthropology at University College London, and Peter Wheeler, Director of Biological and Earth Sciences at John Moores University, Liverpool, suggested humans were able to develop larger and more efficient brains because they also evolved smaller and more effective guts.[18]

Known as the Expensive Tissues Theory, this groundbreaking idea arose from a realisation that the larger the brain, the greater its appetite for energy in the form of glucose. Assuming you are fairly inactive, the energy provided by one fifth of all the food you consume goes entirely to fuelling your brain. Aiello and Wheeler asked themselves what special features humans possess that enable them to deliver more glucose to the brain than other animals. One possibility might have been that we burn up energy at an exceptionally fast rate. It turns out, however, that a person at rest burns energy at almost exactly the rate predicted for any primate of our body weight. This ruled out the possibility that our big brains are powered by inordinate amounts of energy passing through the body.[19]

This left only one possible explanation. The metabolic cost of brain tissue is approximately 240 kcal/kg per day (at rest), which is similar to other organs such as the liver (200 kcal/kg per day), and around half that of the heart (440 kcal/kg per day). Yet, as we have seen, the brain accounts for approximately 20% of total energy usage in adult humans, compared to 8–10% in primates and 3–5% in all other mammals. The immense energetic demands of the brain, therefore, can only be met by a corresponding reduction in the amount being consumed by other organs in the body.

Among primates, the size of most organs is determined by inescapable physiological rules. A species whose body weighs twice that of another, for example, requires a heart weighing almost twice as much. The sole organ that could be reduced in size without significant cost to health was the digestive tract – provided it could be supplied with foods rich in calories and nutrients.

As Figure 4.1 shows, while our liver, kidneys and heart are of the size we would expect in an animal of our size, the brain is three times larger and the gut about 60% smaller than expected.[20] Reducing the size of our gut allowed us to expand our brain, while keeping the total weight of our five key organs at approximately 4,400 grams, which is almost exactly in line with the 4,452 grams we would expect them to be in a primate of our size.

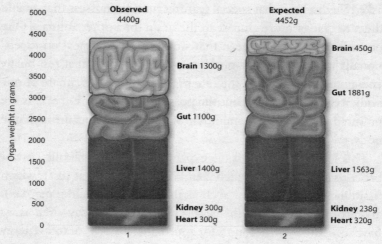

Figure 4.1 Observed vs. expected weight of different organs in the body. As this graph shows, while human brains are considerably larger than expected, the weight of our guts is well below what would be expected. The observed weight of liver, kidneys and heart are all roughly in line with expectations. From Aiello, L. C. and Wheeler, P. (1995), with permission.

'The relationship between relative brain size and relative gut size is perhaps not unexpected,' comment Leslie Aiello and Peter Wheeler. 'Of all the expensive tissues only the brain and the gastro-intestinal tract have a significant latitude to vary in size in relation to the overall body size of the animal.'[21]

The Better the Diet the Smaller the Gut

The size of an animal's gut depends on the quality of its diet. Watch cows grazing in a field and you'll quickly realise that's almost all they ever do. For six hours a day they chomp and chew grass, hay and any other vegetation that comes their way.

While this diet provides a certain amount of carbohydrates and a little protein, the fact that their food is full of fibre, lignin (the organic substance which supports plant and wood cells) and even some silica, makes it difficult to digest. For every 24 calories a cow consumes, less than 1 calorie (0.95) is converted into new tissue, while 15 calories are expelled as faeces, urine and gas.

To achieve even this meagre energy conversion cows, in common with other grass-eating ruminants, possess four stomachs. Here the food can be partially digested before being returned to the mouth for further chewing, after which it is returned to the stomachs once more. It takes a great deal of grass to provide ruminants with the nourishment they need, which explains why most grazing animals are fairly large – they have to be able to accommodate their complex digestive systems, which can include up to 170 feet of intestine. This compares to an average length of 22 feet 6 inches (6.9 metres) in a human male and 23 feet 4 inches (7.1 metres) in a female.[22]

Aiello and Wheeler estimated the number of calories a species is able to save by having a small gut, and showed that the number nicely matched the extra cost of a larger brain. They concluded that because primates have smaller intestines and therefore expend less energy fuelling them, they are able to allocate more to power their bigger brains than other animals. Humans have proportionally the largest brains and smallest digestive systems of any primate. The brain's energy requirements, and how we 'learned' to get greater nutrient value from food, sets up the next critical phase of understanding the brain in the context of obesity.

The Value of a Bigger Brain

Some evolutionary biologists claim that the pressure for large brains, and the greater intelligence these make possible, arises

chiefly from the need to outwit social rivals. The late Richard
Alexander, from the University of Michigan, argued that because
brainpower is critical for planning raids and winning battles,
possessing a higher intellect could lead to greater success during
the long history of human conflict.[23]

Another theory proposed that species hunting over great
distances required superior brainpower to create mental maps of
their territories. But while it is true that early hunter-gatherers
covered distances significantly greater than those of apes, there is
no discernible correlation between range size and brain size.[24]

A more definitely observable advantage of possessing a larger
brain is that it enables more complex social groups to develop. It
allows individuals to develop both 'social intelligence' and a
'Theory of Mind' – that is, the ability to recognise that other
individuals think much as they do, which leads to feelings of
empathy. With the finding that more sophisticated primates with
bigger brains and a greater amount of tissue in the neocortex
(frontal regions of the brain that contribute to long-term decision-
making) live in larger groups, have more close relationships and
use strategic planning more effectively than smaller brained
primates, evolutionary psychologist Robin Dunbar proposed this
'social brain' hypothesis.[25]

But leaving aside the undoubted benefits of the increased intelligence which a larger brain affords, one of the most important
questions is how we obtained sufficient energy to grow more brain
tissue in the first place. Brain tissue is extremely 'expensive'; it has
enormous energy demands, and therefore needs a lot of fuel to
keep it functioning. Where did that energy come from?

Two million years ago it is likely our diet comprised almost
exclusively plant material – fruits, rhizomes, nuts, potatoes and
other root vegetables, seed pods and tree gums – and perhaps
meat obtained through random scavenging or occasional
hunting. It was an existence characterised by brief periods of

prosperity interspersed with times of extreme hardship, and even famine.[26]

Because the length of time an animal can survive without food increases with body size, large animals are more tolerant of variation in food availability. Seals and whales, for example, are capable of surviving off their own body reserves for extended periods, and can transfer a significant amount of energy to their offspring via their milk. Where humans are concerned, an advantage of our comparatively larger body size lies in the ability to consume low-quality 'fallback' foods such as potatoes and root vegetables, bulbs and corns. Meat is higher quality but harder to come by. So digestive adaptability enabled humans to intersperse grazing and consumption of plant foods with consumption of animal tissue, giving them multiple potential sources of energy. Such flexibility would have aided survival when hunting trips failed and undoubtedly helped to provide the demanding brain with the energy it needed.

However, many researchers believe there was another critical catalyst for the growth of our larger brain – at some point in our evolutionary history, we humans learned to cook.

The Discovery of Cooking

In 1942, nutritionist Victor Lindlahr published a book entitled *You Are What You Eat*. The phrase quickly became a popular piece of folk wisdom. It is, however, not entirely accurate. Rather, we are what we ingest and digest, with the emphasis less on ingestion than digestion.[27] The way our bodies process the nutrients we put into them is absolutely crucial.

People who eat only uncooked fruits and vegetables have to consume many more of them to prevent weight loss, since they can fully digest only a portion of what they ingest. Cooking softens hard-to-digest foods such as root vegetables, making them easier to chew and to absorb nutrients from. Heat also breaks up the long molecular chains present in many toxins, whilst also killing bacteria and other parasites in meat.

Exactly when and how mankind, the only species to cook its food, first worked out how to do so remains uncertain. One idea, which must in the absence of firm evidence be regarded as highly speculative, is that, between 700,000 and 800,000 years ago, a wildfire accidentally cooked some animals. Scavenging people that came across these found them so tasty they adopted the procedure for themselves.[28]

It is possible, but again mere speculation, that the first cooking involved simply tossing the raw meat onto an open fire. Some time later spit-roasting likely became the favoured method; one can imagine hunters returning with their prey on the end of a spear and holding it over an open fire, turning the meat to ensure it was cooked evenly. The invention of sharp tools, first in stone and later in metal, enabled them to cut meat into smaller pieces to cook it faster. Food might also have been boiled, using the shells of large molluscs or turtles as vessels; clay pots were only created in about 5000 BCE.

'Cooking increased the value of our food. It changed our bodies, our brains, our use of time, and our social lives,' says Richard Wrangham in his book *Catching Fire: How Cooking Made Us Human*. He believes that the 'transformative moment that gave rise to the genus *Homo*, one of the great transitions in the history of life, stemmed from the control of fire and the advent of cooked meals.'[29]

The most widely accepted evidence for mankind's first controlled use of fire comes from Gesher Benot Ya'aqov in Israel. At this waterlogged site on the shores of Lake Hula, in the northern Dead Sea Rift, archaeologists identified early hearths, on which willow, poplar, ash and wild olive had been burned.

Evidence was also found for the roasting of oats, wild grapevine and barley as well as meat. The animals consumed included horses, deer, rhino, hippo and birds. Bones were unearthed showing cut marks and patterns of breakage indicating the extraction of marrow. On the basis of this evidence, Professor Naama Goren-Inbar and her colleagues from the Institute of Archaeology at Hebrew University concluded that: 'The hominins who frequented the shores of the lake for over 100,000 years knew how to use fire and exercised that knowledge repeatedly throughout much of the . . . period. The domestication of fire by hominins surely led to dramatic changes in behavior connected with diet, defence and social interaction.'[30]

By increasing the efficiency with which food could be digested, cooking freed our early ancestors from the need to spend so much time foraging, chewing, and digesting. The higher-quality diet that resulted from cooking enabled proto-humans to absorb energy from food more efficiently. This meant they were able to get by with a smaller gut, which allowed an increase in the size and sophistication of their brains.

The Thrifty Gene

While our distant ancestors might have lived in deprivation, always unsure where their next meal was coming from, today, in the developed world at least, consumers enjoy unlimited access to brain fuel in the form of HED (high energy-dense) foods. Indeed, the chief challenge in the twenty-first century is to *avoid* consuming too much glucose. So what happens when we no longer need our guts to work so effectively, when we have more fuel than our primitive ancestors could have ever dreamed was possible?

Half a century ago, James Neel, a professor in the department of Human Genetics at the University of Michigan Medical School, published a controversial paper entitled, 'Diabetes Mellitus: A

"Thrifty" Genotype Rendered Detrimental by "Progress"?'[31] In it he sought to address a medical paradox that had long puzzled health workers. As mentioned in Chapter 3, diabetes is a serious condition, and therefore impairs evolutionary success by making sufferers less likely to reproduce. This seems contrary to basic Darwinian principles – why should such harmful genes have been favoured by natural selection? The answer, suggests Neel in his 'thrifty genotype' hypothesis, is that they once aided survival.

'During the first 99 per cent or more of man's life on earth, while he existed as a hunter and gatherer, it was often feast or famine,' he points out. 'Periods of gorging alternated with periods of greatly reduced food intake. The individual whose pancreatic responses minimized post-prandial glycosuria (the passing of glucose in the urine) might have, during a period of starvation, an extra pound of adipose reserve.'[32]

So Neel argues that insulin resistance, which allows for huge blood sugar spikes and ultimately favours fat storage, would actually have been beneficial for our distant ancestors. It would limit the liver's ability to absorb and metabolise glucose, ultimately leading to increased fat storage. This additional fat would be immensely useful as a source of energy during periods of famine. However, in more luxurious circumstances this same trait contributes to an increased risk of developing Type II diabetes, which is what we are seeing happen today. No matter how helpful these genes might have been during the Palaeolithic era, they are distinctly unhelpful in conjunction with a twenty-first century diet.

When Things Go Wrong – the Barker Hypothesis

As we have explained, between birth and twelve months babies become increasingly fat. Later, between one and five, when children grow taller, they no longer require large stores of adipose tissue, and gradually become slimmer. Then, around the age of six, the so called 'adiposity rebound' occurs. The child starts putting on weight again, laying down fat deposits once more. Exactly when this rebound occurs is of critical importance to the child's health in later life. For reasons that are still not fully understood, low weight gain during the early years triggers an early adiposity rebound. And an early adiposity rebound is a strong predictor of both Type II diabetes and obesity in adulthood.[33]

More than twenty years ago, David Barker, Professor of Clinical Epidemiology at the University of Southampton, and one of the twentieth century's most influential epidemiologists, demonstrated that babies with a below-average birth weight are at greater risk of developing a wide range of diseases later in life.[34] He was not only concerned with the effect of malnutrition due to poverty on babies, but to that of any diet unbalanced in macronutrients and deficient in micronutrients. Diets, in fact, very similar to those followed by millions of people worldwide today.

This 'foetal programming hypothesis', which the *British Medical Journal* later named the 'Barker Hypothesis', transformed thinking about the causes of obesity and obesity-related disorders such as coronary heart disease, stroke, hypertension and diabetes.[35] Challenging the notion these can be explained by a combination of bad genes and unhealthy lifestyle, Dr Barker argued that the root of all these health problems is to be found in the womb. He claimed it is the nutritional environment in which foetus and infant develop which permanently 'programs' metabolism and determines the 'pathologies of old age'.

'The nourishment a baby receives from its mother, and its exposure to infection after birth', he says, 'determine its susceptibility to chronic disease in later life.'[36] The very start of life is the critical developmental window on which the individual's future health and wellbeing depend.

To understand why things go wrong, let's focus on the role of insulin as the baby develops in the womb.

Insulin before Birth

Manufactured by beta cells in the pancreas, insulin is the sole hormone responsible for energy storage in vertebrates, including humans. Only the brain and liver do not require it in order to utilise glucose. Because of insulin's unique role, a great many metabolic problems arise if there are any difficulties over its release.

Additionally, insulin controls the rate at which a foetus develops by matching speed of growth to the availability of food. If the mother is poorly nourished, so too is the foetus. In this situation an automatic survival mechanism kicks in. Muscle growth is sacrificed, since it is of relatively low importance to the developing baby. By keeping more sugar in the blood, rather than allowing it to be stored in the muscles, nature strives to protect the developing brain of the baby. After a while, in conditions of nutritional scarcity, this cautious approach to sugar management – the 'thrifty genotype' discussed earlier – becomes 'hard-wired' into the brain and so persists throughout life.

If, following birth, children who have developed in such conditions find themselves in an environment where food is freely available, they have an increased risk of developing diabetes, either because their bodies are incapable of producing sufficient insulin, or because their tissues do not respond to it in the normal manner.

A reduced ability to produce insulin, combined with a need for it in larger quantities because the body is less responsive to it, makes it impossible for the individual to maintain blood sugar levels within normal limits. The individual has developed what doctors term 'insulin resistance'. If a person is insulin resistant, their muscle, fat and liver cells do not respond appropriately to the hormone. The body therefore needs to produce more insulin in the pancreas to aid the transfer of glucose to those cells. If it is unable to meet the demand, excess glucose accumulates in the bloodstream. If this glucose is not used by the cells, it is converted into fat. Thus, by compromising the body's ability to utilise glucose, insulin resistance can cause weight gain.

A previously advantageous strategy for safeguarding the brain is now transformed into a life-threatening liability. The blood becomes flooded with sugar and weight gain occurs, which then increases the body's resistance to insulin. When the body has too much adipose tissue and too much blood glucose over a prolonged period, diabetes develops.[37] So we have good reason to think that the diet of a mother during pregnancy can have long-term adverse effects on a child's ability to control their weight, and serious implications regarding their likelihood of developing diabetes.

Fat Mothers Produce Fat Babies

It's a fact that obese parents tend to have obese children. While only 10% of children with slender parents become obese, approximately 40% of those with one obese parent do. Where both parents are significantly overweight, this proportion rises to 70%.[38]

The question posed by Professor Alexandra Logue is: how much 'is due to eating or exercise habits transmitted environmentally from parents to children, and how much is due to genes transmitted from parents to children?'[39]

To investigate the link between the mother's weight and her unborn child's risk of obesity, Professor Neena Modi and her colleagues, from the Department of Medicine at Imperial College London, used Magnetic Resonance Imaging to scan over a hundred newborn babies as they slept. We will have more to say about this powerful technique for imaging both body and brain in Chapter 7.

The scans enabled them to measure the amount of fat in the infant's liver cells as well as the total amount and distribution of fat in their bodies.

What the researchers found surprised and concerned them. Liver cell fat and total fat, especially around the abdomen, increased across the entire range of children in correlation with their mothers' BMI; the more overweight the mother, the fatter her baby.[40] This study produced the first evidence of biological changes which, combined with an unhealthy lifestyle, could put these infants' health in jeopardy as they grew older.

'Fatter women have fatter babies and there is more fat in the babies' livers', Neena Modi concluded. 'If these effects persist through childhood and beyond, they could put the child at risk of lifelong metabolic health problems. There is growing evidence that a baby's development before birth has a major impact on their health in later life. This means that the prevention of obesity needs to begin in the womb.'[41] In short, the results of Professor Modi's work bear out the Barker Hypothesis.

The accumulation of excess adipose tissue, fat, is not the result of greed or laziness. Rather, since glucose is the primary fuel of the brain, it may be the case that the intense psychological pressure and stress we feel on a daily basis causes the brain to cry out for more fuel. Thus, excess adipose tissue is perhaps better viewed as an unintended consequence of the brain's chronic demand for more glucose. The next problem, of course, is that excess adipose tissue makes further demands on the brain by transmitting a signal that the body is in fact hungry, which is something we examine with the discovery of the hormone leptin, covered in the next chapter.

In this chapter we have explained some of the reasons that we strongly believe that obesity is not down to greed or laziness.[42] Whilst poor diet choices and insufficient exercise play an important role in weight gain, ultimately it occurs in response to demands from the brain. The brain is, in turn, affected by hormones circulating in the bloodstream. To understand what these hormones are, and how they exert their influence over consumption, we must investigate the secret life of fat.

CHAPTER 5

The Secret Life of Fat

'Fat is one of the most fascinating organs out there . . . we are
only now beginning to understand fat.'
Dr Aaron Cypess, Harvard Medical School[1]

Think about fat and what image comes to mind?

Perhaps a lump of shiny white lard, the orange grease soaked
into the cardboard of a discarded pizza box, or maybe the
congealing sludge you pushed to the side of your plate in disgust
as a child?

When it comes to humans, fat is often seen as the outward
manifestation of gluttony and greed; as Nature's punishment for
the sin of overindulgence, visited upon the guilty in the shape of
beer bellies, muffin tops, love handles, dimpled thighs, bulging
waists and flabby buttocks. It is something to agonise over and
'burn off' by means of rigorous dieting and vigorous exercising.

To regard fat in this way is, however, to do an exquisitely complex
and vital organ a considerable injustice. Over the past thirty years,
scientists have come to appreciate that it is far from being simply
an inert storehouse for energy. Like the heart, brain, liver, and
kidneys, fat is a living, metabolically active organ.[2] It produces and
secretes a variety of powerful hormones into the bloodstream and
is in constant communication with every other part of the body.
What fat 'says' during these exchanges, and how its messages are
received, is no less important than lifestyle choices in determining
whether we stay slender, put on weight or become obese.

Fat, or adipose tissue, consists primarily of cells called adipocytes (also known as lipocytes), which specialise in storing energy. There are also a smaller number of other cell types in fat, including those from which adipocytes evolve, called preadipocytes and fibroblasts, which form the basic structure of cells and tissues. There are two types of adipocytes, named white and brown fat due to their respective appearances. Of these, White Adipose Tissue (WAT) is present in the greatest quantities in a human body, and plays the more significant role in weight gain.

WAT – Fat as a Hormone Factory

In the late 1980s, Pentti Siiteri, a biochemist at the University of California, San Francisco, known to friends and colleagues as 'Finn', identified adipose tissue as an endocrine organ.[3] Derived from the Greek *endo*, meaning 'inside' and *krinein* meaning 'secrete', such systems comprise the glands, cells and tissues which metabolise and exude a wide variety of hormones directly into the bloodstream.

The interactions between hormones and their target cells might be likened to software programs. By regulating metabolism and modulating reactions to different foods they run our bodily processes. While we cannot *consciously* control hormonal responses, for example by instructing the pancreas to release more insulin, we can and do exert significant influence over them via our behaviour; by the amount and intensity of the exercise we take, the quality of our sleep, the amount of stress we experience and by what we eat.

Hormones produced by white adipose tissue (WAT) play a critical role in metabolism by balancing energy needs and controlling the cell's sensitivity to insulin.[4] Its colour, and hence its name, derives from the fact that each cell consists of a large drop of pure white lipid (liquid fat made up of triglycerides and cholesterol ester) surrounded by a membrane. The average adult has around 30 billion white fat cells, accounting for around 20% of male and 25% of female body weight. This fat gobbet takes up so much space that the cell's nucleus is squeezed into a thin rim around the periphery (See Figure 5.1).

Figure 5.1. A white fat cell. The amount of space taken up by the reservoir of fat (triglycerides and cholesterol ester) means the nucleus is pushed to the periphery.

In a person of normal weight, WAT cells have a diameter of around 0.1 micron, but are capable of expanding by up to four times their normal size when storing excess fat. If they become too large they can divide and so increase the number present in the body. This, however, only seems to occur during childhood and adolescence. It's not that overweight adults have more fat cells; theirs are just larger.

White Adipose Tissue serves four key functions:

1) It provides thermal insulation, enabling the body to maintain core temperature within very close limits.

2) It produces hormones that control metabolism and some aspects of sexual activity.

3) It stores excess calories and releases them when hunger strikes. It does this in response to insulin, produced by beta cells in a region of the pancreas called the Islets of Langerhans.[5]

4) It plays a critical role in inflammation, the body's response to stress. Inflammation of fat cells is one of the key stages in the development of Type II diabetes, which has made the role of

inflammation and fat a key priority in obesity and diabetes research.[6]

WAT is found in various locations around the body, but precisely where it is deposited varies according to sex and age. Women tend to accumulate it on their hips, legs, and buttocks, while men store it mainly around their waist. Researchers term the resultant pear shape in women as gynoid and the resultant apple shape in men as android. There is also a third shape, termed androgynous, where the overall distribution of fat is neither typically male nor typically female but somewhere between the two, leading to a more rectangular shape.

Excessive WAT around the abdomen, termed visceral fat, poses a far greater risk to health than fat stored beneath the skin. While hormones secreted subcutaneously circulate to every part of the body, those produced by visceral fat cells flow into the portal artery and travel directly to the liver, where they can cause serious damage.

Unlike in cowboy movies, where the man in white is always the good guy, in our body it is white fat that is the villain – when either deficient or present in excess, that is. Too much white fat is associated with a wide range of diseases, and associated with metabolic syndrome. The term metabolic syndrome (MetS) describes a cluster of health problems that emerge in response to the body's inability to cope with excessive fat tissue, problems such as diabetes, hypertension, heart disease and chronic inflammation. Yet, a lack of adipose tissue is not desirable either; although extremely rare, lipodystrophy is a condition marked by an individual's inability to store fat appropriately. Curiously, the same problems associated with obesity ensue; an individual who suffers from this condition often has problems with insulin resistance and a fatty liver.[7]

Adipocytes produce a variety of hormones, referred to collectively as adipokines or adipocytokines. *Adeps* is the Latin for fat, *kytos* the Greek for container (in this case, a cell), and *kinein* Greek meaning 'to move'. Metabolic syndrome develops when adipokine secretion runs out of control, leading to such obesity-linked problems as inflammation, elevated lipid levels (hyperlipidemia) and vascular disease – all risk factors in heart disease – and Type II diabetes, previously summarised. Three of the most important adipokines are resistin, adiponectin and peptin.

A Closer Look at Resistin

First identified in 2001 by endocrinologist Mitchell Lazar and his team at the University of Pennsylvania School of Medicine, resistin's main role is in regulating metabolism.[8] Levels of resistin in the blood have been shown to rise in conjunction with an increase in white adipose tissue around the waist, and to decline when weight is lost.[9] Increased levels are also associated with a rise in diabetes risk; its name refers to a suggested link to insulin resistance. Currently, however, this idea has not received general acceptance in the scientific community.

Resistin has been demonstrated to cause high levels of bad cholesterol (low-density lipoprotein or LDL) in the human liver and also to degrade its LDL receptors. As a result, the liver is less able to clear 'bad' cholesterol from the body. Resistin also accelerates the accumulation of LDL in the arteries while adversely affecting the impact of statins, the main cholesterol-reducing drug used in the treatment and prevention of cardiovascular disease.

A Closer Look at Adiponectin

Discovered by four independent groups of researchers, in the mid-1990s, adiponectin is known to the scientific community by a variety of names – Acrp 30 (adipocyte complement related protein of 30 kDa), apM1 (adipose most abundant gene transcript 1), adipoQ and GBP28 (gelatin binding protein). Manufactured by a gene located on a chromosome which is reported to have a link to obesity (chromosome 3q27), its function is to regulate fat and glucose metabolism.[10]Adiponectin levels vary in inverse relation to obesity, which has naturally caused significant interest among obesity researchers.

Adiponectin plays a role in the diseases present in the metabolic syndrome. Low levels of adiponectin in the blood, which are found in around a quarter of the population, are associated with insulin resistance, Type II diabetes, coronary artery disease, vascular disease, hypertension and obesity.[11]

Koji Ohashi and his colleagues at Osaka University's Graduate School of Medicine and the Molecular Cardiology Whitaker Cardiovascular Institute at Boston University found that in mice, low adiponectin levels were associated with a rise in systolic blood pressure. When levels were topped up, the pressure declined. 'These data suggest,' reports Koji Ohashi, 'that . . . low production of adiponectin might relate to the pathophysiology of hypertension.'[12]

Mother's milk contains large amounts of adiponectin, which might explain the finding that babies who are breastfed have a reduced risk of obesity later in life. We will also have more to say about the crucial importance of human milk in the next chapter where we describe its vital role in the growth of gut bacteria.

Because adiponectin acts directly on the brain, its effects are rapid and profound. When, in 2004, researchers at the University of Pennsylvania School of Medicine injected it into mice, the rodents lost weight, not because their appetites had been curbed but because their metabolic rate had been increased. 'The animal burns off more calories, so over time loses weight,' wrote Rexford Ahima, then Assistant Professor of Medicine at Penn. Diabetes Center. 'Here we have another fat hormone that can cause weight loss but without affecting intake.'[13]

While studying diabetes, Professor Philipp Scherer and his colleagues at the University of Texas in Dallas worked with a group of mice that had been genetically engineered to secrete about three times the normal amount of adiponectin, limiting the amount reabsorbed by the body. As a result these animals ate so voraciously they almost doubled their weight within just a few months. The one thing they singularly failed to do, despite their obesity, was become diabetic.

'The continual firing of adiponectin generated a "starvation signal" from fat that says it is ready to store more energy,' explains Professor Scherer. 'The mice became what may be the world's fattest mice, but they have normal fasting glucose levels and glucose tolerance.'[14]

The million-dollar question was – why?

The answer came when Philipp Scherer and his team examined the way body fat had been deposited in the mice. While they found an abundance stored directly beneath their skin, there was very little in organs such as the liver. This led them to speculate that the unusual distribution might explain why the grossly overweight animals remained in such good health. Excessive fat in the liver can make the organ less sensitive to insulin, which as we have already discussed can lead to diabetes. 'This indicates,' writes Scherer, 'that the inability to appropriately expand fat mass in times of overeating may be an underlying cause of insulin resistance, diabetes and cardiovascular disease.'

This discovery also suggested that the subcutaneous fat cells in individuals with low adiponectin levels failed to signal that they were ready to accept more fat. Adiponectin may also play a vital role in directing where fat is stored and this, in turn, plays a crucial role in the development of pathologies associated with metabolic syndrome. The mice that secreted high levels of adiponectin had fat reserves in all places in their body apart from their organs. Thus, adiponectin may protect an individual from cumulating fat in locations more hazardous to health, such as the liver, heart and muscle tissues. It is not just how much fat is being stored but where it is deposited that makes a difference between health and disease. 'It is a little bit like real estate,' remarks Scherer, 'location, location, location.'[15]

A Closer Look at Leptin

The story of leptin begins with the world's fattest mouse and ends with a discovery that forever changed the way scientists view obesity. It started in 1958, at the Jackson Laboratory in Bar Harbor, Maine. This low-storey, red brick biomedical research centre can fairly be described as the kingdom of the mouse. Not only has the lab sequenced the entire mouse genome, it also produces more than 5,000 strains of genetically modified mice which are used in medical research all around the world. However, of the 5,000 strains known to scientists as the Jackson, or JAX, mouse, one strain in particular is among the most intensively researched and academically famous rodents of all time, having been used to gain a better understanding of obesity and diabetes.

It was in 1958 that a 27-year-old biochemist named Douglas Coleman joined the Jackson Laboratory. He was intending to stay for just a couple of years, but was to remain there for his entire professional career. The laboratory, he later explained, 'provided a rich environment, including world-class animal models of disease, interactive colleagues, and a backyard that included the stunning beauty of Acadia National Park.'[16]

Coleman took his first degree, in chemistry, at McMaster University, where he also met his future wife: 'the only girl to graduate in chemistry in the Class of 1954', then went on to complete a PhD in biochemistry at the University of Wisconsin. His early researches were into muscular dystrophy and the newly emerging field of mammalian biochemical genetics. He had little or no interest, at that time, in studying obesity. Then, in 1965, something unexpected happened in the mouse colony at the Jackson Laboratory, a chance event which was to change his professional life forever and set in motion research that was to occupy him for much of the next thirty years.

A massively obese mouse was born.

When Coleman first started working there, the Jackson Laboratory was already home to what was then the world's most important murine strains, known as the obese mutant mouse. Named the ob/ob mouse ('ob' stands for obese), it was the result of a chance genetic mutation. Seven years later a second spontaneous mutation occurred producing a second strain of obese mice. Like the ob/ob mouse, these animals ate voraciously and grew rapidly but, unlike the ob/ob mice, they also suffered from Type II diabetes. This, as we saw in Chapter 4, is caused either by the pancreas failing to manufacture sufficient insulin or because the cells of the body have become resistant to it. Obesity in this second strain was found to have arisen as the result of two defective copies of a gene, which researchers dubbed 'db' for diabetes, leading to them being named db/db mice.[17] The ob/ob and the db/db mice appeared outwardly identical to an observer, in terms of both looks and behaviour.

Douglas was given the task of discovering what the precise biological difference was between these outwardly similar mice. The fact that both strains were morbidly obese led him to wonder whether these similarities might be due to a common factor circulating in their blood. To test this theory he performed an operation, known as parabiosis, in which he surgically attached a db/db to a normal mouse in such a way that they shared the same blood supply. Within a week, despite the animals being given all they wanted to eat, the normal mouse was dead. It had perished not as a result of surgical complications but from starvation. Its Siamese twin partner, the db/db mouse, had not only survived but also grown even fatter. Douglas repeated the experiment on numerous occasions with exactly the same outcome. Each time, after about seven days, the blood glucose concentrations in the normal mice had fallen to starvation levels.

'The normal mice not only consistently lacked food in their stomachs and food remnants in their intestines but also had no detectable glycogen in their livers,' Douglas Coleman recalls. 'In marked contrast, the diabetes partners consistently retained elevated blood sugar concentrations, and their stomachs and intestines were distended with food and food residues.'[18]

When he performed the same surgery on two genetically normal mice, however, both remained well and active for months afterwards. At which point he experienced what he later called his 'Eureka!' moment.

'These results led me to conclude that the db/db mouse produced a blood-borne satiety factor so powerful that it could induce the normal partner to starve to death.'[19] There was something in the blood of the db/db mice that told the bodies of the normal mice that they were full and required no extra energy when, after a few days, precisely the reverse was true.[20]

Despite these clear results, many of his colleagues, as well as researchers in other laboratories, refused to accept them. They remained firmly wedded to the prevailing doctrine that obesity was a consequence of lifestyle, not hormones; that it was overwhelmingly due to the fact that overweight people ate unhealthily and exercised infrequently, if at all. Over the next few years, however, more and more researchers came around to the view that obesity also had an underlying physiological cause.

The hunt for the satiety factor then became a race. Drug companies knew that, if it could be found and marketed as the ultimate slimming pill, it would make a fortune. After all, the US weight-loss industry alone produces annual revenues in excess of $20 billion.[21]

Over the next few years many possible candidates were suggested as the satiety factor (cholecystokinin, somatostatin and pancreatic polypeptide, for example). However, after rigorous experimentation, each was discarded.

Leptin and the Accidental Scientist

The breakthrough was finally made not in Douglas Coleman's Bar Harbor laboratory, but a short flight down the East Coast in New York City. At 5.30 on the morning of 8 May 1994, Jeffrey Friedman, a forty-year-old molecular geneticist at Rockefeller University, identified the factor and the gene that produced it.

'It was astonishingly beautiful', he says, talking about the X-ray film that identified the gene involved. It is an image that now takes pride of place on the wall of his office.[22]

Jeffrey Friedman never intended to be a researcher. He wanted, and expected, to follow his father into medicine. Born in Orlando, he spent most of his childhood in Long Island, New York, where his father practised as a physician. 'The world of my parents' parents was essentially the world of immigrants,' he says. 'And in that world, you did whatever you could to get your feet on the ground. And in my family, the highest form of achievement was being a doctor.'

So, young Jeffrey set out to study medicine. Despite doing extremely well at school, this did not prove an easy ambition to realise. After receiving rejections from seven colleges, he was finally accepted on a six-year programme at Rensselaer Polytechnic Institute and Albany Medical College of Union University in upstate New York. An MD by the time he was twenty-two, Jeffrey then began a residency at Albany Medical Center Hospital, with no concrete plans about how to spend a gap year between leaving that job and embarking on a fellowship at the Brigham and Women's Hospital in Boston.

'One of my professors thought I might like research,' Jeffrey remembers. 'Why he thought I might have some particular aptitude, I can't really tell. He said, "I have this friend at Rockefeller, why don't you go spend a year with her and see if you like research?" I didn't know what else I was going to do. My mother thought I should go spend the year as a ship's doctor!"[23]

Rejecting her advice, Jeffrey joined the laboratory headed by Professor Mary Kreek and almost immediately fell in love with her research into the way molecular biology could provide a road map for understanding behaviour.

'That was 1981 and it was beginning to be evident that molecular biology was going to have a big impact, so instead of going to the Brigham for a fellowship, I abandoned medicine and decided to get a PhD with Jim Darnell . . . one of the leaders in molecular biology.'[24]

Though he did not know it, Friedman was giving himself all the preparation he needed to make one of the most important discoveries in the history of obesity – and biological – research. Soon after establishing his own laboratory at Rockefeller in 1986, Friedman started to hunt for the elusive satiety factor. It proved a long and difficult search, requiring many years of meticulous experimentation. 'Today, it would take about a week to do what we did in the lab,' Friedman told us. 'Back then, it was a process of about eight years.'[25]

He described to us how, late one Saturday night, he was looking at blots with RNA (ribonucleic acid)* from the fat tissue of normal and mutant mice. Blot analysis is a method for transferring biological material, such as protein, DNA or RNA onto a cell membrane. The chemical reaction is registered on photographic film, providing scientists with a way to determine the gene expression levels of particular genes.

---

* RNA and DNA (deoxyribonucleic acid) are the nucleic acids which, together with proteins and carbohydrates, make up the three major macromolecules that are essential for life.

After preparing a blot while investigating a new, unknown hormone, he went home but, unable to sleep, got up at around five in the morning, returned to his bench and developed the blot. What he saw made him almost pass out in surprise.

'When I looked at the data, I immediately knew that we had cloned *ob*, that is, the gene which had made the first JAX mouse so fat in the first place. When I saw it, I was in the darkroom, and I pulled up the film and looked at it under the light and got weak-kneed. I sort of fell backwards against the wall. This gene was in the right region of the chromosome, it was fat specific, and its expression was altered in two independent strains of ob mice.

'Before this, we didn't know where ob would be expressed – and while fat was one of the tissues I considered, in principle the gene could have been expressed in any specialized cell type anywhere that had no obvious relationship to fat. But, on the other hand, seeing a gene in the right region expressed exclusively in the fat . . . that gets your attention. We saw RNA in the normal animal – but a ten times increase in fat from the mutant animal? At first that might be puzzling, but that's when I knew we had identified a gene. The fact that one animal didn't make any RNA, and the other showed an increase in RNA, meant there had to be a change in genetics.'

At six in the morning Friedman telephoned his wife and, wild with excitement, told her 'We did it!'[26]

Later that day, he met up with friends for a drink: 'We opened a bottle of champagne, and I told them, "I think this is going to be pretty big."' It was the moment every scientist dreams of. But what Jeffrey Friedman could not then know was just *how* important his discovery was to prove.

'It's the truth to say it's the closest thing to a religious experience I've ever had,' Jeffrey Friedman told us. 'A lot of scientists will explain things to be beautiful. In our case it's a few blobs on an X-ray, but its clarity was incredible. At first, it seems odd to use an aesthetic description for a scientific reality, but it's really the only way to explain that moment of discovery. I had a result that was aesthetically beautiful, it described how nature has learned to monitor too many calories that lead to fat accumulation; the answer is through making a hormone.'

Having pinpointed the gene responsible, his next task was to give it a name. Initially he intended to call it simply the 'obesity gene' but the Nobel Prize-winning French endocrinologist Roger Guillemin challenged his choice as inappropriate. After visiting Friedman at the laboratory, to learn about the discovery at first hand, Guillemin returned to Paris and wrote Friedman a letter: 'I really liked what you had to say, but I have one quibble: you refer to these as obesity genes, but I think they are lean genes because the normal allele keeps you thin. But calling them lean genes sounds awkward. The nicest sounding root for 'thin' is from Greek, so I propose you call ob and db 'lepto-genes'."[27]

Friedman played around with this suggestion and finally decided to name the hormone he had identified as the satiety factor 'leptin'. It was a discovery that many scientists regard as one of the most important advances of the twentieth century, and was the result of years of painstaking research and some extremely obese mice.

The discovery of leptin has allowed a far more humane and accurate portrayal of the disease of obesity. It is a condition in which the victims are physiologically – rather than exclusively psychologically – unable to control the amount they eat.

How Leptin Works

Leptin, 80% of which is secreted by subcutaneous white adipose tissue, has been dubbed the 'hunger hormone', and with good reason.[28] Its function is to constantly monitor and regulate energy levels, a task it accomplishes in three main ways:

By counteracting the effects of a substance called neuropeptide Y, a potent appetite stimulus secreted by certain cells in the gut and by the hypothalamus, an organ buried deep within the brain.

By counteracting the effects of anandamide, another appetite stimulant.

By promoting the synthesis of a hormone with the tongue-twisting name of α-Melanocyte-stimulating hormone, which acts as an appetite suppressant.

Levels of leptin in the blood are directly related to the severity of obesity. When present in sufficient quantities and when brain and gut are receptive to its signals, leptin turns off the desire to eat the moment that energy needs have been met. In everyday parlance, our appetite ceases once we have eaten enough to sustain normal bodily function.

However, when leptin levels are too low and/or the brain and gut are, for some reason, unable to receive its signals, this 'off' button is never pressed, our hunger is never satisfied and our desire to carry on eating becomes uncontrolled and relentless. The situation is worsened by the fact that leptin is manufactured in inverse proportion to the amount of fat already in store. In other words, the fatter an individual, the less leptin they secrete and the more urgently and persistently their brain demands food.

'When peripheral signals – such as leptin and insulin – are not released,' says Nora Volkow, Director of the National Institute on Drug Abuse, 'or your brain becomes tolerant to them, you don't have a mechanism to counter the drive to eat . . . It's like driving a car without brakes.'[29]

Leptin doesn't just control energy intake and body weight. It also plays a role in such diverse processes as the immune response, bone development and cell growth. It even influences the time taken for wounds to heal.[30]

There is also evidence that it may affect the ease with which we breathe. Leptin encourages inflammation, and is found in higher levels among asthmatics, regardless of the extent of their obesity. Perhaps most surprisingly of all, leptin has a significant influence over fertility, sexual desire and the onset of puberty. To understand why a hormone whose primary function is the control of appetite should be involved in this most intimate of human behaviours we need to step back through time.

As we explained in Chapter 4, the lives of our earliest ancestors were marked by periods of famine interspersed with occasional feasts. When the crops were harvested, when fruit had ripened on the trees and when hunters returned with a fresh kill, the whole tribe could gorge themselves. But when food was scarce and there was insufficient energy available to meet all the body's needs, the brain had to allocate what resources it had to the most essential activities: breathing, pumping blood, filtering urine and digesting what little food was available. With survival at stake, leptin was used to slow down or close down all activities that were not immediately essential, such as reproduction, sex drive, puberty, menstruation and even fighting off disease.

Thin, undernourished girls and boys, for example, reach puberty at a later age than do well-nourished girls and boys. Indeed, under extreme circumstances they may remain prepubescent for years. In women, menstruation will cease entirely under starvation conditions.

So the effects leptin has on the chemistry of the body are wide-ranging and significant. But it was its relationship with appetite which interested many people first and foremost. It seemed like it might be the magic bullet that would banish obesity without any need for dieting.

After all, if weight problems are largely caused by a leptin deficiency, then surely introducing it into an overweight person, perhaps by a pill or injection, would immediately diminish their appetite, allowing the excess pounds to drop away. Sadly, but perhaps not unexpectedly, given the complexity of human metabolism, it is not that simple.

The $20 Million Gamble that Failed

Two years after Jeffrey Friedman's discovery of leptin in mice, the first proof that the hormone also played an important role in humans was found; in 1997 researchers reported on their findings from studying two morbidly obese children.[31]

After being injected with leptin, these youngsters reduced their eating significantly and rapidly lost weight. Excited by the apparent potential of leptin in the treatment of obesity, Amgen – a biotech company – paid Rockefeller University $20 million to license the hormone. They gambled that, with increasing numbers of people becoming overweight or obese, such a treatment would prove highly lucrative, as well as offering a significant improvement in public health. A large clinical trial was started, in which leptin was injected into overweight adults. The results were disappointing; although a small group of obese patients did lose significant amounts of weight, the overall magnitude of the effect was minimal.[32] Shortly afterwards, Amgen suspended research into leptin for treating human obesity. Their $20 million gamble had failed.

Why Leptin Failed

If Jeffrey Friedman was disappointed by the outcome of the Amgen trials he was not surprised: 'Even before Leptin was tested in obese patients, we knew from animal studies that this hormone was not likely to be a panacea for every obese patient and that the response seen in ob/ob mice wasn't going to be the typical case for obese humans. Leptin levels are elevated in obese humans, suggesting that obesity is often associated with leptin resistance and raising the possibility that increasing already high levels was going to be of arguable benefit.'[33]

Clearly, obesity is not simply an issue of leptin deficiency in the blood plasma, if injecting the hormone doesn't actually change anything and if there are elevated levels of leptin in obese humans anyway. Rather, the issue must be leptin resistance. This occurs when receptors in the brain either no longer respond to signals from the hormone or do so with insufficient sensitivity.

But what causes this problem in the first place?

While the situation is still not entirely clear, the finger of blame seems to point firmly in the direction of diet. Animals in the wild, which do not usually stray far from their evolutionary 'diets', have properly functioning metabolisms in which the leptin pathway is preserved. Most modern humans, by contrast, now have a diet significantly different from what was eaten in even the fairly recent past. The result is that millions are metabolically challenged by seriously disrupted leptin pathways.

This being the case, the key to developing a successful anti-obesity treatment involving leptin is likely to lie in finding ways of increasing the sensitivity of receptors in the brain to leptin, or by preventing resistance.

And the fast-growing field of personalised medicine, where treatments are developed to match a patient's unique genetic make-up, may have something to contribute. It could allow doctors to identify which individuals will respond positively to injections of the hormone, like the small subgroup in the original Amgen trials. At the same time, research is being undertaken into other agents that might be combined with leptin to enhance its effectiveness with a wider group of people.[34]

Brown Adipose Tissue (BAT)

There is a type of adipose tissue, known as brown fat, or brown adipose tissue (BAT), which has been observed to do something truly remarkable in laboratory mice: it protects the animals from becoming obese by burning more calories the more they eat. In human terms this is the equivalent of losing weight by over-indulging on French fries! It's definitely worth looking at BAT in greater depth.

The distinctive colour of brown fat cells is due to the fact they contain a uniquely large number of organelles called mitochondria, which appear brown because they contain iron (see Figure 5.2). The function of mitochondria is to convert oxygen and food energy into cell energy or, in the case of brown fat, into heat. While brown fat occurs in rodents throughout their life, scientists once believed it was only present in humans during infancy and early childhood, when it helps maintain body temperature. The accepted medical wisdom was that by adulthood so little remained in the body as to have no significance.[35] This is now known to be false.

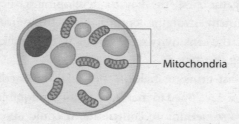

Figure 5.2. A brown fat cell – six mitochondria,
which appear brown because they contain iron, can be
seen among the globules of fat.

The Hunt for Adult Brown Fat

The world-renowned Joslin Diabetes Center is located in Boston's leafy Joslin Park, a brief stroll from the city's 'Emerald Necklace', a series of parks and waterways covering more than a thousand acres. It was here, in 2009, that researchers announced a potentially game-changing discovery in the understanding and treatment of obesity. They had analysed almost two thousand body scans in the search for brown fat in adults. The results were published in the prestigious *New England Journal of Medicine*.[36]

They were striking. 'Not only did we find active brown fat in adult humans, we found important differences in the amount of brown fat based on a variety of factors such as age, glucose levels and, most importantly, level of obesity,' explains Dr Aaron Cypess of Harvard Medical School, a research associate at the centre.

'What is of particular interest is that individuals who were overweight or obese were less likely to have substantial amounts of brown fat,' says Ronald Kahn, Head of the Section on Obesity and Hormone Action at Harvard Medical School, another member of the research team. 'Likewise, patients taking beta-blockers and patients who were older were also less likely to have active brown fat. For example, individuals both over age 64 and with high BMI scores were six times less likely to have substantial amounts of brown fat.'[37]

Brown fat was detected mainly in younger, leaner subjects in the regions of the neck and collarbone and was found much less frequently in overweight or obese individuals. Women had detectable regions of brown fat twice as often as men. Although the amount of brown fat found did not represent a high proportion of total body fat, being only 8% in women and 3% in men, the researchers also noted that the real amount was likely to be considerably higher; scans are capable of detecting brown fat cells of a certain size and activity, but almost certainly fail to identify smaller, less active deposits.

This matters, since even small amounts of brown fat could play a significant role in combating obesity. Somebody weighing 65 kg (143 lb), for example, might have only 65 to 80 grams (7 to 28 oz) of brown fat, compared to between 9 kg (19 lb) and 14 kg (30 lb) of white fat. However, if maximally stimulated, this could be capable of burning off at least 100–200 calories a day, which would be sufficient to lose almost half a kilogram (1 lb) of (white) fat each week.[38]

There is evidence that certain activities – such as spending time in the cold – can somehow activate brown fat, leading to increased caloric burn and a shift in other metabolic and hormonal functions. In a recent investigation into energy expenditure of men with and without brown fat, it was shown that exposure to cold activated brown fat and subsequently led to an increase in resting metabolism of 15% among those men with the extra reserves of brown fat, compared to those who did not have it.[39]

There is currently a burgeoning field of research dedicated exclusively to whether brown fat can be cultivated, and if so, how. While it continues to be an important therapeutic target, researchers are hesitant to proclaim the benefits: 'Research papers say cold will "turn on" brown fat and cold will change the hormones that regulate body weight, but it doesn't give you that next step that says sleeping in the cold will make you a fashion model,' explained Steven Smith MD, one of the researchers at Sanford-Burnham Medical Research Institute, where significant research is being conducted in this field.[40]

By finally recognising fat's key role as an active organ, medical researchers have helped to demolish decades of dogma surrounding the causes of obesity. While sadly society still has quite a long way to go before it catches up with the facts, we now know with certainty that obesity is far from solely being linked to lifestyle choices. It is a serious and potentially life-threatening medical condition and should not be stigmatised as simply the just reward for gluttony and weakness of will.

CHAPTER 6

Gut Reactions

'The relationship between the mind and the body is more complex than has previously been presumed . . . the body too might be influencing the mind to a greater degree than has previously been recognized.'
Iris Hung & Aparna Labroo[1]

The 81-year-old woman was sick and dying. Dr Max Nieuwdorp was, he admits, 'young and naïve'.[2] But he was also determined to save the elderly patient's life.

She had been admitted to Amsterdam's Academic Medical Centre with complications following a urinary tract infection. After the antibiotics used to treat her infection had annihilated her colon's natural microbial population, *Clostridium difficile*, a far less beneficial bacterium, took up residence in her guts. When Dr Nieuwdorp first met his patient she was unable to eat, running a high fever, had bedsores, an inflamed bowel and chronic diarrhoea. Repeated doses of the antibiotic vancomycin, the treatment of choice for *C. difficile*, had proved ineffective. As increasingly happens, the notorious pathogen – which kills 300 patients a day in the US alone – had developed a resistance to what is generally an effective treatment.[3] With the most powerful weapon in their pharmaceutical armoury useless, there was little her physicians could do other than make their patient's last few hours as comfortable as possible.

But Max, recently appointed to a residency in internal medicine at the hospital, refused to give up. The treatment he had in mind, however, was so unorthodox that his supervisor, Joep Bartelsman, first thought the young doctor had to be joking.

Before we describe what that treatment involved, let's take a step back to review a relationship which has intrigued philosophers and scientists for centuries: the interactions between what are generally referred to as the mind and the body, but which we prefer to describe as the brain and the body.*

Despite the seemingly obvious link between a person's psychological wellbeing and their physiological health, the medical community has generally shunned this holistic approach. Today, however, we are starting to see a resurgence of interest in the role this relationship plays in both health and weight gain. We can look to physiologists and neuroendocrinologists to provide insights into what is known as the brain–gut axis; to uncover the lines of communication that occur between these two organs in order to help us understand how each affects, and is affected by, the actions of the other. Many believe that advances here will shed light on the current obesity pandemic.

This brings us back to our young doctor and his proposed treatment for the dying 81-year-old.

'A bizarre and disgusting treatment'

'I want,' Max Nieuwdorp told a surprised Joep Bartelsman, 'to administer a faecal transplant.' In non-medical terms he was proposing to pump the stool of a healthy person – her son – into the lady's guts![4]

In fact this wasn't a new idea; ancient Chinese healers had been known to employ a mixture they termed 'yellow soup', hot water

---

* The reason we make this distinction is a simple one. No one knows what the 'mind' is. Even its existence as something somehow distinct from the body has come to be doubted by many neuroscientists, ourselves included.

mixed with faeces, to carry out a similar procedure. Even in more modern times faecal matter transplant had been employed with some success – Ben Eiseman and his Colorado-based team of surgeons used the technique to treat a select group of patients suffering from colitis. Most were cured within two weeks.[5]

Yet, it has been Drs Bartelsman and Nieuwdorp who have re-established this technique within academic circles, and pushing for its inclusion as a treatment in a broader, more accessible context. While recognising that a lot of people, including many doctors, would regard this as a bizarre and perhaps a disgusting treatment, Dr Bartelsman gave his approval – after all, there was nothing to lose. Both doctors knew that however distasteful it might seem, it was the old lady's last and only chance of survival. Using a blender, the medics mixed her son's stool with a saline solution and then infused the resultant liquid into their patient by means of an enema. That done, all they could do was cross their fingers and wait to see whether the highly unorthodox procedure would succeed.

It did. And beyond their best expectations. Three days later the lady walked unaided out of the hospital. Her life saved by something, which, under normal circumstances, would have been flushed away down the lavatory.

To understand what this has to do with obesity, we need to go back over half a century and review the pioneering work of Ben Eiseman, Chief of Surgery at Denver General Hospital.

In 1958 Eiseman published a report in the journal *Surgery* on four patients suffering from 'pseudomembranous colitis', a painful, highly debilitating and potentially fatal inflammation of the colon associated with *C. difficile*. After all other treatments had proved ineffective, Eiseman decided to try an approach which had actually first been carried out in fourth-century China. Using the method that Max Nieuwdorp would duplicate years later, he obtained faecal matter from donors, blended it with saline and transplanted it into his patients. All made rapid recoveries.[6]

Despite Eiseman's success, it was not until the mid 1980s that Australian gastroenterologist, Professor Thomas Borody, of The Centre for Digestive Diseases in Sydney, Australia, brought the idea back into mainstream medicine. Polish-born Borody, who had immigrated to Australia with his parents as a child, was in a quandary about how best to treat a female patient with ulcerative colitis. This is a painful and distressing inflammation of the colon, resulting in abdominal pain, diarrhoea and weight loss. Some doctors believe it to be an autoimmune condition, that is, one in which the immune system, mistaking harmless bacteria as a threat, attacks the colon causing it to become inflamed. There is, at present, no known cure and all doctors can do is try to relieve the symptoms.

Having tried the recommended drugs, which included steroids, without success, Dr Borody was uncertain what to do next. Thumbing through the medical literature in search of guidance, he chanced upon Eiseman's paper and decided to try it: 'I looked at the method and I kind of made up the rest of it,' he recalled. Once again, this unorthodox treatment produced an almost instant cure.

Although faecal transplants were clearly effective, most doctors at that time dismissed them as both repulsive and dangerous.[7]

'I was initially ostracized,' Thomas Borody recalled in 2011: 'A professor of medicine named me on television as being a charlatan for doing faecal transplants and he had no idea of the science behind it . . . even now my colleagues would avoid talking about this or meeting me at conferences, although this is changing.'[8]

Despite the treatment's acknowledged success, even some of the physicians who practise it admit to revulsion: 'Part of me has not overcome that feeling of disgust with human waste. I think it's universal,' admits Alex Khoruts, Belarussian-born gastroenterologist and immunologist at the University of Minnesota. 'We're supposed to avoid the stuff . . . it's essentially like the elephant in the room for the gastroenterologist. We talk about all the other parts of the digestive tract, but we're so ignorant about this component . . . so our level of knowledge hardly exceeds that of a fifth-grader who just says, exactly as you said, "Eeewww."[9]

Even as late as 2006, when Max Nieuwdorp and Josep Bartelsman decided to treat another six *C. difficile* patients using the same procedure, their embarrassment at doing so was such that they waited until the other doctors had gone to lunch before carrying it out. Today, Nieuwdorp and Bartelsman have taken the lead in publishing articles and empirical reviews on the procedure. The success rate for faecal bacteriotherapy, to give the procedure its medical name, has been remarkable. Nine out of ten *C. difficile* patients are reported to be free from their infection after only one transplant. Faecal transplants have also been reported to lead to improvements in sufferers of Parkinson's disease, multiple sclerosis, Crohn's disease, ulcerative colitis, acne, Type II diabetes, muscular dystrophy and perhaps even obesity.[10]

So what is it that brings about such astonishing outcomes? How can there be any benefit in transplanting into one person the discarded waste of another?

To answer these questions we need to go deep inside the human body and examine the actions of the multitude of bacteria living inside our guts. It is on the health of this unseen and unsung army of microscopic helpers that our own health, sometimes our lives, depends.

Microbes by the Billion

In the late seventeenth century, a Dutchman named Antony van Leeuwenhoek acquired an interest in microscopy. Although he came from a relatively humble background, van Leeuwenhoek taught himself to grind lenses and build microscopes. During a long life (he died aged 91), van Leeuwenhoek constructed around 500 of these simple devices. While they were really no more than powerful magnifying glasses, they afforded a glimpse into a previously invisible and unexplored world.

Using one of his microscopes to examine plaque scraped from his teeth, van Leeuwenhoek discovered: 'Little living animalcules, very prettily a-moving.'[11]

More than a century later, Ferdinand Cohn, a German botanist, named these 'animalcules' 'bacteria', from the Greek bakterion, meaning a 'short rod', since that is what some he observed under his microscope resembled. By 1853, Cohn had gone on to identify three types of these microorganisms: bacteria, with their short, rod-like structures, bacilli (longer rods) and spirilla (spiral forms).

Bacteria have existed on earth almost since the dawn of time. The oldest known fossils, almost 3.5 billion years old, are those of bacteria-like organisms. There are around 40 million bacterial cells in a teaspoon full of soil, and a million in half an egg-cup of fresh water. The 5×10^{30} (that's 10 followed by thirty zeros, an almost inconceivably large number) bacteria on Earth form a biomass greater than that of all the plants and animals combined. Bacteria give yogurt its tang. They make sourdough bread sour and beer alcoholic. They break down dead organic matter and contribute to diseases, ranging from allergies and asthma to cancer and heart disease. Crucially, bacteria also produce molecules that kill other bacteria. It is in this ability of 'good' (i.e. health-giving) bacteria to kill and replace 'bad' bacteria in the human gut that the efficacy of faecal transplants lie. Seventy per cent of our immune activity takes place within the gut. Unfortunately, antibiotics cannot tell the difference between 'good' and 'bad' bacteria; when successful they just kill them all indiscriminately, leading to potentially serious consequences for health.

The Power and Promise of Poo

'We are,' says Thomas Borody '10 percent human, 90 percent poo.'[12] Distasteful as this may seem, the point he is making is in regard to microorganisms; our body plays host to a vast and diverse community of microorganisms that live on us and in us. On our hands, feet, face, chest, abdomen, feet and genitals, in the mucosal linings of our mouth, in stomach and intestines. Known collectively as the 'microbiome', it is estimated to consist of more than 1,000 different species and over 7,000 different strains.[13]

The microbiome is becoming such an important priority in health research that the American National Institute of Health has recently spent over $173 million in order to describe and characterise the role of the human microbiome and microbiota.[14]

While microbiome refers to the *total* number of organisms and their genetic material, the term 'microbiota' is used to describe the population present in specific bodily systems, such as the intestines. Our intestinal microbiota, which weighs approximately a kilogram, or a little over 2 lb, contains some 100 trillion microorganisms at densities of around 10^{12} cells per millilitre. Ninety-nine per cent of these microorganisms are bacteria; between 500 and 1,000 bacterial species are estimated to reside within the colon, the final loop of intestine. What is sometimes referred to as our 'gut flora' also includes fungi, protozoa and archaea. The latter, single-celled prokaryotes (meaning they have no nucleus) are very similar to bacteria, for which they were once mistaken, but are now known to be a different type of organism. That said, little else has been established about their functions inside our bodies.

Once regarded as mere tenants, microscopic stowaways hitching a ride on their hosts, recent research has shown a far more intimate connection exists between us and these intestinal microorganisms.[15] They help us digest fibrous materials and harvest calories; as much as 15% of the energy used by the average adult could not be obtained without their assistance. Around 95% of a body's serotonin, a neurotransmitter linked, among other things, to mood and depression, is manufactured in the intestinal microbiota. They also protect their hosts from marauding and harmful bacteria. Laboratory animals that have been engineered to have a complete absence of intestinal microbiota have no functioning immune system. As a result, their white blood cells remain dormant, their intestines fail to develop correctly, their hearts are shrunken and they suffer other health difficulties. So many, so varied and so important are the activities of these bacteria that the intestinal microbiota can justifiably be described as our body's forgotten organ.[16]

In return for their assistance, the microbes gain from us a warm, moist home that provides protection and nourishment. To get some idea of the extent of the presence of microorganisms in our gut, consider this: two thirds of faecal matter's dry mass consists of microorganisms.

'The human body can be viewed as an ecosystem,' Elizabeth Costello of Stafford University School of Medicine asserts, 'and human health can be construed as a product of ecosystem services delivered in part by the microbiota.'[17]

The Making of the Microbiome

An unborn baby is the only truly 'human' human on earth. Up to the moment of birth, we are all 100 per cent human. From that point on, 90% of the cells in our body are microbes, as we have over 100 trillion microbes all over the body. We become, in a sense, 10% human and 90% microorganism. Our microbiome contains around 150 times more genes than our human genome. The microbe cells are much smaller than our own, meaning the microbiome only weighs in at around 200 grams, but its effect on our body and behaviour is profound.[18]

'Think of it as a hulking instruction manual compared to a single page to-do list,' suggests Moises Velasquez-Manoff, author of *An Epidemic of Absence: A New Way of Understanding Allergies and Autoimmune Diseases*.[19]

Except in rare cases where microbes invade the amniotic cavity, the foetus remains in a completely sterile environment and free from any of the microorganisms among which life outside the womb will be spent. The moment the membranes of the amniotic sac rupture and the mother's waters break, all this changes abruptly and dramatically. Even as they travel down and out of the birth canal, vaginally delivered infants receive a strong input of vaginal and possibly urogenital and faecal microbiota.[20]

Apart from their route of entry into the world (a vaginal or Caesarean birth), a person's microbiome will be influenced by such factors as whether they are breast or bottle fed;[21] their diet, hygiene, social and sexual behaviours; genetics; interaction with other children, parents, pets, the home environment; and whether or not they receive antibiotics. In short by virtually anything and everything with which they come into contact.

Within a few years, a bustling community of between 500 and 1,000 species will have been formed, with some having been passed down from one generation to the next – like family heirlooms. By their first birthday, an infant will be the possessor of a unique and idiosyncratic gut microbiota that sets the stage for health later in life.[22]

The Importance of Mother's Milk

'Blood, urine, saliva, and spinal fluid, these are the human bodily fluids most explored by scientists over the decades', points out science writer Trisha Gura. 'Yet any woman who has ever nursed a newborn will cite a major omission: breast milk.'[23]

It is only in the past few years that what microbial ecologist David Mills of the University of California has described as 'a genius fluid that was outrageously understudied' has really started to be appreciated.[24]

Previously the majority of doctors regarded breast milk as simply a source of food for the rapidly growing infant. Doubtless many parents only ever think of it as such, even now. But while human milk certainly contains a rich mixture of fats, proteins and sugars, in many ways more vital than the nourishment it provides is the protection it affords. Breast milk is replete with immune cells, stem cells for regeneration, and thousands of bioactive molecules. These protect the infant against infection, prevent inflammation, strengthen the immune system, promote organ development and help form the microbiome.

The key here is a bacterium with the tongue-twisting name of *Bifido-bacterium longum biovar infantis* which dominates the microbiome during the first months of life, making up as much as 90% of its bacterial content. After weaning, that proportion has fallen to just 3% of the adult microbiome. How this bacteria gets into the baby's gut initially is not yet known – it may be as a result of swallowing amniotic fluid while passing through the birth canal, or even through the milk itself.

Fascinatingly, contained within human milk are substances known as human milk oligosaccharides (HMOs) – some 200 have been identified – whose purpose is to feed *not* the baby but the bacterium. 'Mother is recruiting another life form to babysit her baby,' comments food chemist Bruce German.[25]

This 'babysitting' role consists of protecting the neonate against a host of potentially deadly infections. One way it does this is by consuming the available food and so starving harmful microbes, such as salmonella, listeria and campylobacter. Another of its functions is to provide food, in the form of short-chain fatty acids (SCFAs), for other beneficial bacteria.

These continue to change and develop for the next fifteen or sixteen years, achieving maximum diversity and stability at adolescence and remaining stable until the later stages of life.[26] It is only as we age that our microbiota starts to decline, becoming less diverse and stable, predisposing the elderly to infections such as *C. difficile*. The number and variety of bacteria increase exponentially the further one travels along the digestive tract, with the final section – the five-foot-long, greyish-purple colon – harbouring the majority of intestinal microbiota. The bacteria here are largely what are termed 'obligate anaerobes', meaning they will perish if exposed to oxygen. They maintain a high population density by deriving nutrients from food in the gut and from the intestinal epithelial lining. The processes through which bacteria derive food for themselves produce by-products that in turn affect our body chemistry, which may lead to enhanced satiety. For example butyric acid – found in milk, butter and Parmesan cheese – is produced by bacterial fermentation of dietary fibres, and may serve as an energy source and lead to feelings of fullness.[27]

Helpful Fellow Travellers

While each individual microbiome is completely unique, we do all share a broadly similar profile, 90% of which comprises bacteroidetes, firmicutes and proteobacteria. Thus it has been suggested there is a 'core' microbiome. Bacteroidetes are rod-shaped bacteria that, in addition to living in our mouth, guts and skin, are also found in such environments as the soil, sea water and on the skin of other animals.

Firmicutes (from the Latin *firmus* (strong) and *cutis* (skin)) are a phylum that includes some notable pathogens, or disease-causing bacteria, and have a unique relationship with obesity. Organisms with higher concentrations of firmicutes are often heavier and have greater fat mass. They are either round (cocci) or rod-like structures (bacilli). Because they produce endospores, which are resistant to desiccation, firmicutes are able to survive in extreme environmental conditions, and thus can inhabit the gut with relative ease.[28]

Proteobacteria were named after the Greek sea god Proteus because, like him, they are able to assume a great number of shapes. This major phylum includes various pathogens such as salmonella and vibrio (which can cause gastroenteritis, septicaemia and cholera). It also includes *Helicobacter pylori*; this lives in the stomach, where it helps regulate levels of the hydrochloric acid used to break down food, though it can also cause peptic ulcers, chronic gastritis and even stomach cancer.[29] We will be revisiting *Helicobacter pylori* a little later in this chapter when we talk about the role of antibiotics in disrupting and deregulating the microbiota.

What Does the Intestinal Microbiota Do?

'We depend on our microbial partners for essential services,' says Ruth Ley from the Department of Microbiology at Cornell University, 'such as energy harvest from food, its detoxification, a supply of vitamins, and protection against harmful invaders.'[30]

Gut flora defends us against pathogens by strengthening the wall of the colon to help prevent bad bacteria from entering and by producing antimicrobial substances such as the antibody immunoglobulin A (Iga). As mentioned above in connection with breast milk, by competing for food these beneficial bacteria also starve unwelcome invaders out of existence. There is also strong evidence to indicate that they play a pivotal role in determining whether or not someone becomes obese, a theory to which we will return later in the chapter.

And crucially our gut flora help keep us nourished by unlocking energy from the fermentation of undigested carbohydrates and the absorption of short-chain fatty acids. They play a role in synthesising vitamins B and K as well as metabolising such essential substances as bile and sterols. The gut microbiota produce a dizzying array of enzymes that enhance digestion. Without them we would not otherwise be capable of processing difficult-to-digest foods such as complex carbohydrates, fat and protein; we would not be able to absorb them properly and they would therefore end up being stored in the body, contributing to the development of excess fat.

Several factors, such as infection, disease and diet can adversely affect the microbiome.[31] Researchers have found, for example, that switching from a low-fat diet to one high in fats and sugars can alter the structure of the microbiota within a single day. So too does our increasing use of antibiotics. Given the integral role of these bacteria in our digestive processes, it's hardly surprising that this can have significant effects.

Turning Off the Hunger Hormone

Abnormal patterns of microbiota are consistently seen in those who suffer from obesity, and its associated illnesses.[32] One way in which microbiota contribute to gut health and hunger pangs may be through the modulation of the hormone 'ghrelin'. The stomach produces two hormones responsible for controlling hunger. One is leptin, which we described in the previous chapter; the other is ghrelin (from the Indo-European root *ghre* meaning 'to grow').

Ghrelin switches our appetite on while, as we have explained, leptin turns it off again. Mounting evidence suggests that ghrelin interacts with a particular bacterium found in the gut called *Helicobacter pylori*; eradicating these bacteria from the gut causes metabolic disturbances.

'When you wake up in the morning and you're hungry it's because your ghrelin levels are high,' Martin Blaser, Professor of Internal Medicine and Microbiology at New York University explained in a paper in 2005. 'The hormone is telling you to eat. After you eat breakfast, ghrelin goes down.'[33] Problems arise when antibiotics are taken to treat an infection. These destroy not only the harmful but also the helpful bacteria in the gut, including *H. pylori*.

The result of losing *H. pylori* as collateral damage in this way is often an increase in weight – without *H. pylori* to modulate ghrelin concentrations in both the stomach and blood plasma, appetite can run out of control. In one study, ninety-two US veterans treated with antibiotics gained significant amounts of weight compared to their counterparts who did not use antibiotics.[34]

Given the frequency with which we now use antibiotics in medicine, *H. pylori* is actually being eradicated from our bodies at a shocking rate. Two or three generations ago, 80% of Americans had *H. pylori* present in their guts. Today it can be found in fewer than 6% of American children, so while not essential to human survival, it is important within the context of obesity.

'We now are more than 60 years into the antibiotic era and, in developed countries, children regularly receive multiple courses of antibiotics for various ailments, especially *otitis media* (ear infection),' said Blaser in 2005. 'If each course of antibiotics eradicated *H. pylori* in 5 to 20% of cases, the cumulative effect of childhood antibiotic regimens would remove a substantial proportion of colonisations.'[35]

So one of the key factors in the obesity pandemic may be changes, partly due to antibiotic use, in intestinal bacteria. One of the reasons people overeat may be because their depleted micro-biota have difficulty in regulating their appetites or absorbing sufficient energy from food.[36]

In light of its relationship with ghrelin, *H. pylori* has become another target in the quest to understand the exponential rise of obesity. But how can something happening in our guts influence the sensation of hunger experienced in our brain?

Our Second Brain

Long before scientists had any notion that microorganisms resident in the guts played an essential role in the brain's development and function, they knew that there is, inside our guts, what amounts to a second brain. Known as the Enteric Nervous System (ENS), it contains around a hundred million neurons, about the same number as in the brain of a cat.[37] The Enteric Nervous System forms part of the Autonomic Nervous System or ANS, whose important role in stress and anxiety will be described in Chapter 9.

Our two brains, the one in the head and the other in the guts, are in constant two-way communication. Together they regulate digestion to precisely meet the body's varying energy demands. As a result anything that affects our 'gut' brain also influences our 'head' brain, and vice versa.

The major neural communications route is along the tenth cranial nerve, also known as the vagus or, less often, the pneumo-gastric nerve. Vagus, from the Latin for 'wandering', is well chosen – it is the longest of the twelve cranial nerves; after emerging from the base of the skull, it meanders down to the major internal organs such as the heart, kidneys, uterus (also known as the viscera) and also branches to the heart, lungs, larynx, stomach and ears. To the brain it carries information about the digestive system;

from the brain it sends signals that regulate viscera's motility, local blood flow and the secretion of enzymes.

When communications break down (and they are vulnerable to attack from a wide variety of sources) the system becomes 'deregulated', potentially resulting in, among other conditions, inflammation, chronic abdominal pains, eating disorders and obesity.[38] While the importance of the brain–gut axis has been recognised for decades, researchers have only recently realised that the microbiome is also an active and highly-influential contributor to this two-way communication network.[39] The messages it sends to and receives from the brain play a pivotal role in whether or not someone becomes obese.

Obesity and the Microbiota

In the early years of the twenty-first century, Dr Ruth Ley and her colleagues at Washington University were working with the same strain of obese Jackson mice we met in the previous chapter. These animals lack the hormone leptin, which, as we explained, controls the body's 'fat thermostat'. Without it, the mice cannot monitor the amount of fat in their body and quickly become massively obese through overeating.

The team noticed that these mice had 50% *fewer* bacteroidetes and 50% *more* firmicutes than their leaner counterparts. When they examined human gut flora, they found the same imbalance in obese people as in the obese mice. The relative proportion of bacteroidetes *increased* in obese people as they lost weight by eating a low-fat or low-carbohydrate diet, while the firmicutes decreased.[40]

Other researchers reported the same effect in patients following bariatric surgery. As with the mice, the ratio of firmicutes to bacteroidetes decreased. These differences suggest that variations in weight between overweight and undernourished individuals, and how much each eats, may be dictated to a large extent not by their brain but by their microbiota.[41]

The link between microbiota and obesity became even clearer when Ruth Ley and her team studied a special strain of mice, bred to have no microbiota of their own. These intestinal *tabula rasa* mice could eat as much of various fattening foods as they liked without ever putting on significant extra weight. After eight weeks on a diet that was 40% fat, they had put on less than half as much weight as their 'normal' peers, despite eating the same amount of food. When the researchers transplanted the microbiota from fat and lean mice into the germ-free strains, those colonised by bacteria from fat donors packed on far more weight than those paired with lean donors.

To try to discover why shifting the bacterial balances affected body weight, the Washington University researchers next compared the microbiota of fat and lean mice at a genetic level. Samples from fat mice showed much stronger activation of genes that code for carbohydrate-destroying enzymes in the various bacteria, which break down otherwise indigestible starches and sugars. As a result, these mice were able to extract more energy from their food than were lean ones. The bacteria were also manipulating the animals' own genes, triggering biochemical pathways that store fats in the liver and muscles, rather than metabolising them. While these effects are relatively small, the researchers believe that, over the course of months or years, they can result in very large weight fluctuations.[42]

Certain microbial populations secrete enzymes, not encoded in the human genome, that enable calories to be extracted from otherwise indigestible polysaccharides (sugars) in our diet. This results in an increase in bacterial fermentation products, the short-chain fatty acids, which influence various aspects of our metabolism, resulting in the storage of fat in the host. The microbiota has also been shown to influence bile acid production, insulin resistance and inflammation, all of which contribute to metabolic disease.[43]

The fact that – as mentioned above – the composition of the gut microbiota varies between individuals has led some researchers to make the controversial suggestion that there is a 'core gut microbiome' which greatly increases the risk of a person becoming obese. Equally, there may be a specific combination of bacteria that impart optimal health benefits.

A recent study examined the effects of infusing intestinal microbiota from nine lean men into another nine males suffering from both metabolic syndrome and obesity. Six weeks afterwards, those with metabolic syndrome showed significantly improved bacteria profiles (more varied bacteria in the gut) and improved insulin sensitivity profiles. In fact, the dramatic improvement of insulin sensitivity has led researchers to speculate whether more traditional forms of enhancing gut bacteria (i.e. via oral supplements) may have an important role in the future of obesity control medicine.[44]

Success in this area has prompted a surge in research that should lead to greater understanding of our gut health, and indeed our health in general. Metagenomics, the name given to this new and rapidly expanding field of research, uses sophisticated screening techniques to characterise the microbial community at a genome level in health and disease. Over the past few years, huge strides have been made in understanding the composition, diversity and functions of the human microbiome. If researchers are able to identify what constitutes obesity-promoting or obesity-resistant gut flora, such knowledge will open up new doors of opportunity for preventive, diagnostic and therapeutic approaches in disorders of microbiota–brain–gut axis.[45]

Research success in understanding the interactions between genetics, diet and human gut microbiota could lead to ways of reducing or even eliminating obesity through modifying intestinal microbiota by means of diet, prebiotics, or advanced probiotics.

While this field of study is still in its infancy, evidence is emerging which strongly suggests intestinal microbiota also play a key role in the development of various aspects of brain function, including anxiety, emotions, cognition and – even more fascinating – sociability. Well-designed and well-conducted randomised trials are now needed to further assess the clinical potential for treating neurodevelopmental and mood disorders.

There is still a great deal that we don't know about our life-long passengers – the exact ways they sense and respond to their host's condition, the details of how they are passed on or how they are affected by our diet. By answering these questions, scientists could then assess whether actively shifting our bacterial balances could help to stem the worldwide increase in obesity levels.

Of course, gut microbiota are not the whole story behind the current obesity epidemic. We need to understand how they interact with other factors that affect our risk of becoming obese, such as lifestyle choices, marketing and advertising techniques, the availability of HED foods and our genetic make-up.

In the next chapter we will be exploring the relationship between the gut and the brain in greater depth to explain how what goes on in the head can profoundly affect what goes on around the waist.

PART THREE

All in the Mind?
Obesity and Your Brain

'The modern brain is optimally adapted to an ancestral society located in the savannas of East Africa and is not especially well adapted to modern social and demographic conditions.'
Daniel Lord Smail, *On Deep History and the Brain*[1]

CHAPTER 7

How Your Brain Can Make You Obese

'I hold that the brain is the most powerful organ in the human body.'

Hippocrates, *On the Sacred Disease* (400 BCE)[1]

The young woman, waiting somewhat apprehensively in the brain-scanning department, was blonde, smartly dressed and obese. She was one of thirty-five volunteers taking part in a study conducted by psychologist Lauri Tapio Nummenmaa, from Finland's Aalto University School of Science. Nineteen of the women were significantly overweight, with an average BMI of 44, while sixteen were slender with an average BMI of 24. What Nummenmaa and his colleagues wanted to discover was whether their brains responded differently to images of food.[2]

As we explained in Chapter 2, we live in an increasingly obesogenic world – an environment in which there is both an abundance of palatable, often low-cost, food and a multitude of powerfully evocative 'food cues'. Yet, while almost everyone is exposed to these enticements to look at food, smell food, taste food, eat food and think about food, not everyone has a weight problem and currently only a minority are obese. This raises the question of how and why a majority, albeit a decreasing one, are able to remain slim. The answer, many researchers believe, is to be found in the way food cues, in their various guises, are processed by the brain. Using brain-imaging techniques they are gradually casting new light on the differences between the brains of individuals of normal weight and the obese.

In this chapter, we will examine various aspects of eating motivation, and try to provide some insight into the differences between eating to satisfy physiological need (what scientists refer to as 'homeostatic' motivation) and eating for pleasure (what is referred to as 'hedonic' motivation). Since obesity can largely be attributed to eating for pleasure, as opposed to need, a clear understanding of hedonic motivation is necessary in order to understand obesity.

We owe a great deal of what we know in this regard to brain scanning technology, such as the functional magnetic resonance imaging (fMRI) used in Dr Nummenmaa's investigation. Using this kind of equipment we are beginning to tease out the differences between an obese person's brain compared to a lean person's.

A white-coated technician helps the young volunteer onto the 'patient table' of a two-million-dollar magnetic resonance imaging scanner – it is a human-sized tunnel surrounded by a large and extremely powerful magnet.

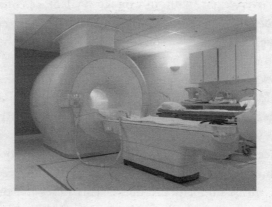

Figure 7.1 The fMRI scanner used by Lauri Nummenmaa in his research. (Photograph courtesy Dr Lauri Nummenmaa and the Department of Biomedical Engineering & Computational Science, Aalto University School of Science, Finland.)

A pillow is placed under the volunteer's knees, to ensure her back remains as flat as possible during the scan, and a white plastic frame is positioned around her head to restrict neck movements. Finally she is handed a 'panic button' with which to summon help if necessary. The technician clicks a switch and the volunteer slides silently into the dark belly of the doughnut-shaped machine. This done, she retreats to the safety of the control room – so powerful is the scanner's magnetic field that, while it is in operation, only the person being scanned is permitted to remain.

The scanner works in the following way: as neurons become more active, their energy consumption increases and they require additional oxygen. This is transported around the body by haemoglobin, a blood protein, which comes in two forms, oxygenated and deoxygenated. When the blood flow to the brain changes, the concentrations of oxygenated and deoxygenated haemoglobin in the blood also change, as oxygen is directed to the areas within the brain that demand it most. It is this Blood Oxygenation Level Dependent activity (BOLD) which fMRI measures. This varies between individuals; for those individuals who show hypersensitivity to food reward, an image of a chocolate cake alone will cause more blood to flow to the brainstem and areas that are responsible for motivating behaviour.[3]

Over a period of forty-five minutes, the machine will take a sequence of thirty images, each depicting a 4-mm thick slice of the volunteer's brain. At the same time she will be shown a variety of colour photographs depicting different types of food. Some will be mouthwateringly delicious treats, such as cakes, pizzas, and strawberries dipped in chocolate. Others will be blander fare like lentils, cabbage and crackers. Despite the fact that for technical reasons no actual food can be sampled, the researchers believe the photographs alone will be sufficient to identify differences in the way the brains of the obese and lean women will respond to the 'anticipatory' rewards they offer.

One important limitation with fMRI is that the timing of any increase in neural activity cannot be determined with any great accuracy. While the scientists can say with precision *where* in the brain the response occurred, they are unable to state precisely *when*.

Despite this constraint, in a crude sense, MRI and fMRI provide a biological map for detecting human motivation. Increased blood flow to a particular area in response to a stimulus is a measurable *physiological* change which depicts a *psychological* phenomenon. If, for example, a subject was shown an image of a delicious food-stuff, greater blood flow to areas of the brain associated with motivation would signify a greater drive to obtain that foodstuff. Measuring blood flow in this way allows scientists to study which stimuli cause the biggest changes in the brain, and to compare the reactions that different individuals (such as obese and lean individuals) have to the same stimuli.[4]

We will return to Dr Nummenmaa's important and revealing study in due course. But first let's look at some of the areas of the brain which play key roles in determining how much we eat, either because they are concerned with our survival, or because they are concerned with pleasure.

The Hypothalamus, Our Body's Thermostat

About the size of a large cashew nut, the hypothalamus is located deep within the brain, just above the brainstem. The hypothalamus could be called the body's thermostat; its function is to maintain homeostatic balance (balance of the body's systems) by coordinating and integrating activities of both the nervous and endocrine system. In this way, the hypothalamus is responsible for regulating blood pressure, body temperature, circadian rhythm, heart rate, immune response, sexual desire, thirst, water balance and, most especially, hunger.

The Brain and Eating: Understanding How We 'Decide' to Eat

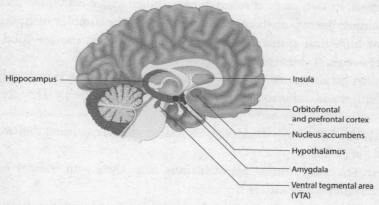

Figure 7.2. Cross-section of the human brain with relevant areas indicated.

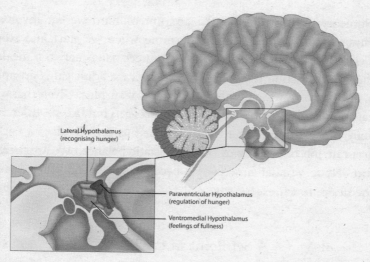

Figure 7.3. The location of the hypothalamus with a close-up showing the three components responsible for hunger and satiety.

Since it is intimately involved with so many crucial functions of the body, including hunger, scientists initially believed failure of hypothalamic functioning to be the major culprit in the obesity crisis. In fact, years of research have shown that it isn't quite that simple: the hypothalamus is but one part of the complex interplay of biological systems relating to our need and love for food. However, it does still have a crucial role to play.

In general, three areas of the hypothalamus are associated with hunger and eating motivation, and are shown in Figure 7.3. They are:

1) The lateral hypothalamus, associated with recognising that we are hungry.
2) The ventromedial hypothalamus, associated with feelings of fullness and satiety.
3) The paraventricular hypothalamus, associated with regulation of hunger.

Homeostatic control over the amount of food we eat involves physiological processes that determine when we start and stop feeling hungry, and which suppress hunger pangs between meals, thereby regulating body weight. The average adult gains around 0.5 kg per year, which is about a 3,500 kcal surplus. Given that we eat about 900,000 kcal per year, this represents only a 0.5% discrepancy, demonstrating that the hypothalamus actually does a fairly accurate job of keeping us in homeostatic balance.[5] We are therefore left to wonder how people become obese, and what drives the desire to eat excessively.

A Set Point in the Brain?

Early studies of obesity were based on the belief that an individual's weight depends on a genetically determined metabolic 'set point'.[6] The theory is that our set point of hypothalamic control guides our eating behaviour at all times. This likens the relationship between our body and brain to a central heating thermostat; if the room temperature falls or rises above a predetermined level, then a thermostat turns the boiler on or off as required. According to set point theory, while weight gain is extremely unlikely to occur in individuals with a naturally low set point, for those unfortunate enough to have inherited a high set point, becoming overweight is unavoidable.

In a lot of ways this seems a sensible theory, as it does help explain why some individuals never seem to have much difficulty in regulating their body weight, while others do. However, it fails to take into account factors such as food choice, metabolic responses to diet, hormone levels, and environment. Moreover, the recent exponential rise in obesity demonstrates the fallacy of our belief that some people are simply 'lucky' with a genetically determined set point. Obesity reflects not only a person's biology, but their environment too, otherwise we would not have seen obesity rates quadruple within the last forty years, across the globe.[7]

'I can eat anything, no matter how fattening, and never put on a pound' is an often repeated boast on the part of those lucky enough to be able to retain a slim figure without much effort. However, as the obesity crisis demonstrates, fewer and fewer people are able to maintain such a lean look. Moreover, a person's perception is everything; 'I can eat anything' may be the case for someone who is thin, but they may only be consuming 1,200 kcal a day. For an obese person trying to diet, they may be overeating without knowing it. Thus, set point theory does not, in fact, provide a comprehensive picture of human weight.

Hypothalamic signalling plays an essential role in guiding our behaviour; it can be held to account for the sensation of when we are 'absolutely starving', the kind of hunger that strikes when we haven't eaten for hours, or are in serious states of energy depletion. Yet, given how accurate the hypothalamus is at keeping us in homeostatic balance, it is clear that eating for pleasure, rather than metabolic need, must be an important component of the obesity crisis. Thus, it is important to look beyond the hypothalamus, to networks within the brain that are responsible for pleasure and reward.[8]

Your Reptilian Brain

Located at the base of the forebrain, and a few millimetres away from the hypothalamus, is a region sometimes dubbed the 'reptilian brain', due to the fact that it is the most basic or primitive part of the whole brain.[9] More correctly identified as the basal ganglia (Figure 7.4), this area actually consists of several distinct regions, the largest of which are the caudate nucleus, putamen and globus pallidus. Present in pairs, with one located in each of the brain's two hemispheres, they are known collectively as the striatum, or 'striped body.'

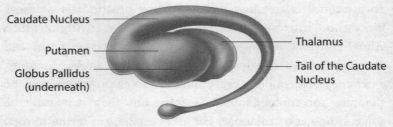

Figure 7.4. The basal ganglia.

Shaped somewhat like the letter 'C', the caudate nucleus has a wide head tapering into a thin tail that curves towards the occipital lobe at the back of the brain (which is responsible for vision). The caudate conveys messages to the frontal lobe, especially to an area just above the eyes known as the orbitofrontal cortex (which we will discuss in more detail shortly). It plays an important role in learning, storing memory, the development and use of language, falling asleep, social behaviour and voluntary movement. Recent research has shown that it responds differently to food in lean individuals than in obese individuals.[10]

The putamen (from the Latin meaning 'shell') is located beneath and behind the front of the caudate. It is involved with two forms of learning: implicit and reinforcement. The former occurs when people acquire knowledge through exposure, for instance by watching TV or studying a textbook. Reinforcement learning involves interacting with the environment and discovering which actions produce the best and most consistently rewarding outcomes. As we will explain later in the chapter, this type of learning plays a significant role in the development of obesity, as obese people may be more vulnerable to the rewarding effects of food. As we describe later in the chapter, eating is a hugely pleasurable experience. Consumption of foods, particularly those high in sugar and fat, elicits a cascade of neurochemical reactions, such as the stimulation of the endogenous opioid, serotonergic, and cannabinoid systems in the brain. All of which are associated with the wonderful, yet fleeting, sensation of eating.[11]

The globus pallidus (Latin: 'pale globe') is located inside the putamen and receives inputs from the caudate and putamen. It sends messages to an area of the brain called the substantia nigra (black substance) which is where dopamine is produced. Dopamine is a neurotransmitter which can produce intensely pleasurable feelings – because of this it is a key player in the brain's reward system, as we shall see in a moment.

Beyond the Reptilian Brain: Where Pleasure and Reward Are Processed

In addition to the hypothalamus and the basal ganglia, there are other parts of the brain involved in eating and overeating, including the amygdala (stress and emotions), insula (sensory integration), and the nucleus accumbens, ventral tegmental area (VTA), orbito-frontal and prefrontal cortex (rewards and self-control).[12] These are shown in Figure 7.2.

The brain has two amygdalas (Greek for almonds, which they resemble in shape and size), one in each hemisphere, which are involved in processing rewards, the desire to eat and detecting the intensity of flavour. They also play a crucial role in stress and anxiety disorders. This is a topic we will deal with in greater detail in Chapter 9 when we discuss the part played by stress and anxiety in overeating.[13]

The hippocampus (Greek for seahorse, named for its resemblance to that animal) is a part of the brain involved in storing memories. It enables us to rapidly recall foods we found pleasurable and rewarding, or those we found unpleasant and would prefer to avoid. Early eating memories can therefore play a key role in determining our tastes and preferences as adults. If, as a child, we came to regard vegetables as disagreeable, for example, this memory is likely to dissuade us from eating them for the remainder of our life.[14]

Another important region is the oval-shaped insula, also known as the Island of Reil. It is essential in determining human behaviour, coordinating autonomic or unconscious bodily functions, in addition to emotional responses.[15] Neuroimaging studies suggest it has anatomically distinct regions, which are concerned with different aspects of taste perception. One part, for example, responds to taste intensity, irrespective of how the diner feels about what is being eaten or drunk, while another mediates emotional responses to the taste.[16]

Interestingly, obese individuals show increased activation in both regions during consumption of food compared to lean individuals, suggesting that they may perceive greater taste intensity as well as experience an increased emotional sensation of reward.[17]

The nucleus accumbens is a part of the brain which becomes especially active whenever people experience intense pleasure, especially from food or sex. It receives information from, and sends signals to, a collection of neurons located on the floor of the midbrain close to the midline; this area is called the ventral tegmental area (VTA), and is involved with thinking, motivation, and the powerful emotions associated with falling in love and experiencing orgasm. On the darker side it plays a role in drug addiction and several psychiatric disorders. It releases dopamine whenever something happens which predicts an immediate reward.[18]

Imagine, for example, that while gazing through the window of a patisserie you notice a mouth-watering cream cake on a display stand on the other side of the glass. And it's the last one in the shop! Anticipating your purchase of that delicious treat, your VTA sends a flood of dopamine coursing through your brain.

Now imagine that as you push open the shop door, the sales assistant hands the cake to another customer. Your desires have been frustrated! You are likely experiencing the sensation of disappointment, and kicking yourself for not acting faster. In such a case it could be argued that the decline in dopamine levels is a response to the over-prediction (and under-delivery) of the cream cake reward. If, however, you purchased and enjoyed the cream cake, an action that met your expectations, then the abrupt dopamine spike would have signalled the under-prediction (and over-delivery) of a reward. If the cake was exceptionally delicious, high in sugar and fat, an even greater sensation of pleasure would ensue. However, if the cake had gone bad, and the cream was spoiled, your reaction would be one of disgust.

This highlights the basic mechanisms by which we learn about food; sensations of reward and punishment are rudimentary Pavlovian mechanisms that shape our behaviour. By understanding this concept, and viewing food in terms of how it 'rewards' or 'punishes' the senses, we can glean greater appreciation of our own relationship with it. Some foods are immediately rewarding, yet have no nutritional value; the high fat salt/sweet taste coaxes our brain into an immediate state of bliss, yet the food does very little for our body. However, some foods that are hugely nutritious taste much less exciting, such as vegetables. Their less sweet, and sometimes slightly bitter taste is something that requires learning and experience to appreciate, yet once this pattern is established one can 'learn' to crave healthy food. The key lies in ensuring the learning phase occurs early, so an individual can reap the benefits of healthy eating habits throughout their entire life, as opposed to having to work to change their food preferences later in life.

Thus for individuals vulnerable to the rewarding effect of food, the constant reminders of it in the modern world, such as ubiquitous advertising, can drive them to constantly search out the rewards they predict from such food cues. This is, in part, what we mean when we say that the world today is an obesogenic environment. In Chapter 11 we will further discuss the powerful role of these cues in the obesity pandemic.

For many years researchers believed obesity reflected hypothalamic dysfunction; however, we now know the essential role that pleasure plays in overeating. Since pleasure circuitry runs throughout the entire brain, and not just within the brainstem, it is likely to have a greater influence on shaping behaviour. Beginning from the nucleus accumbens and extending into the frontal regions of the cortex, overeating cannot solely be attributed to one area within the brain, and is thus a significantly more complex problem to tackle.

While the hypothalamus acts to sense general levels of energy within the body, and the nucleus accumbens and ventral tegmental areas act as 'hedonic hotspots' concerned with pleasure, an outer region of the brain located just above the eyes, the orbitofrontal cortex, is involved in detecting how 'pleasant' a food is and controlling decisions about whether or not to eat it.[19] A study also found that women who fasted were more vulnerable to food-related cues, showing greater activity in frontal areas of the brain that translated to cravings and memories of food in fasting patients.[20]

Finally, there is the prefrontal cortex, which is located just behind the forehead and is the last part of the brain to mature. This is responsible for self-control, judgement and caution and plays a vital role in inhibiting impulsive eating.

Hopefully this whistle-stop tour of various areas of the brain has helped to give a sense of what a complex interplay between the different areas is involved in why and how we eat. There are no completely straightforward answers; rather, neuroscience and technology have allowed a deeper understanding of the networks within the brain, and how they interact with environmental and internal factors to guide 'pleasure-based' decisions to eat.

Dopamine - the 'Gas Pedal' of Pleasure

Dopamine (which also goes by the name 3,4-dihydroxyphenethylamine) is a neurotransmitter which is released from nerve cells in the brain's substantia nigra and ventral tegmental regions, whenever we are engaged in, or anticipate, a pleasurable activity. Dopamine plays a central role in learning and memory, as the larger the release of dopamine is with any given activity, the higher the probability we will remember and repeat that behaviour. Research has shown that dopamine released in the nucleus accumbens is directly implicated in addiction.[21] When it comes to food, enjoying a delicious snack or even just anticipating consuming one, can trigger a tsunami of dopamine which 'hijacks' the reward circuitry of the brain to produce an intensely rewarding experience.

In order for any neurotransmitter that has been produced and released by a nerve cell to affect another, it must first 'lock on' to receptors in it. It is here that one of the main differences between the brains of lean and obese individuals may be found; studies have shown that overweight individuals have *fewer* dopamine receptors than those who are lean.[22] In essence this is because when the dopamine receptors are overstimulated, the brain reacts by pruning them.

Scientists gained an appreciation of the dysfunction of dopaminergic circuitry through the study of addictive behaviour and use of illicit drugs. Through years of addiction-based research, scientists have learned how drugs (particularly amphetamines and stimulants) lead to a surge in dopaminergic activity (which can in turn lead to the pruning of receptors). In more recent years, however, we have learned this can occur with a variety of other substances (including food) and also with behaviours (such as shopping or gambling). For this reason, our definition and understanding of addiction has changed; we are now beginning to view a variety of behaviours as having the potential to be addictive, including eating and overeating.

Researchers believe that once the receptors are pruned, or in scientific jargon 'down-regulated', the individual must consume more of the reward (such as food) to achieve the same amount of pleasure on a subsequent occasion. The 'dose' of food must be stepped up. But more stimulation leads to more receptor depletion, obliging a further increase in the stimulus and meaning that yet more of the reward is needed to achieve the same effect. An addiction of any kind involves this continuing need to achieve the same level of reward as it becomes ever harder to do so.

Neuropharmacologists Luca Pani and Gianluigi Gessa found that: 'The dopaminergic system has remained identical for the last several centuries . . . external conditions which interfere with its physiology have dramatically changed.'[23] In other words, the world in which dopamine developed as part of brain chemistry is vastly different from the one in which it exerts its effects today.

At the core of human nature lies the quest for survival. In order to enhance the probability of survival, we have evolved a pleasure system that reacts to behaviours that the brain and body believe, in order to help us live another day. However, this system can be fooled and abused in what researchers refer to as 'hijacking'; certain behaviours which are pleasurable, such as taking drugs or overeating, actually limit life rather than sustain it. In response to this inbuilt desire for enjoyment, or the scientific term, 'reward', we have created tools, games, social structures, stimulants and foods that are designed to unleash pleasurable sensations. Two centuries ago, we had stricter boundaries regarding the number of friends we could have, the types of drugs we could take and the amount of food we could eat. Today, we've developed technologies and innovations which enable us to circumvent these boundaries.

For some, the drive for pleasure may develop into a pathological need to achieve dopamine's rewarding high, even at the cost of health and wellbeing. Some foods today are specifically engineered to flood the brain's reward centres with as much dopamine as possible, almost inevitably beginning the process of addiction we have described. Obesity demonstrates how exploiting our reward system via food can have disastrous consequences.

Orexin – the New(ish) Kid on the Block

Over the last thirty years, researchers have also come to appreciate the hugely important role played by a tiny group of neurons in the hypothalamus, which were once dismissed as insignificant. It's not hard to see why this was the case – there are only around 20,000 of them, compared to the brain's total of approximately 80 billion neurons.

However, it has now been established that, despite their tiny numbers, they play a key role in a wide range of vital functions, ranging from regulating patterns of sleep and wakefulness, to the experience of pleasure and perception of pain. Their role in obesity demonstrates the tremendously complex interplay of chemical interactions that guide human behaviour and reveal how just a single, tiny deviation in normal brain activity can result in enormous problems for people in everyday life.

There is a particular neuropeptide related to this group of neurons. Neuropeptides communicate between neurons, and are one of the most primitive building blocks for establishing function within an organism. Therefore, they are also implicated in behaviour, as the communication between neurons ultimately leads to all of our actions, thoughts, and perception. This particular neuropeptide was discovered simultaneously by two independent groups and is therefore known as both 'orexin' and 'hypocretin'.

Orexin neurons provide a crucial link between energy balance and the coordination of mood, reward, addiction and arousal. Dysregulation of this infinitesimally tiny brain region produces some terrifying physiological and psychological consequences. These include a tendency to fall asleep in the presence of stress or emotional arousal. Such 'sleep attacks', known as narcolepsy and cataplexy, can be as frustrating as they are dangerous. Narcoleptics may put their life in jeopardy falling into immediate deep sleep, resulting in injuries ranging from bruises to serious trauma. To add to their problems, narcoleptics are often overweight, even when they do not overeat.[24] The laboratory mice whose orexin neurons have been removed exhibit late-onset obesity despite never being allowed to overeat. It therefore seems possible that even non-narcoleptic people could experience problems with orexin and the neurons associated with it that might result in weight issues.

While the fact that the neurons that deal with orexin are located in the hypothalamus is indicative of orexin's role in guiding homeostatically driven hunger, their direct connection with the reward pathway, via the nucleus accumbens, suggests orexin could also be involved in hedonically driven eating. If so, it might be an important focus for trying to minimize the pleasurable feelings associated with either drug abuse, or overeating hyper palatable food.

One possible way in which the discovery of orexin might enable us to control obesity is through the development of an orexin receptor 'antagonist' – a drug that would block the uptake of orexin. In mice on a high-fat diet this has been found to suppress weight gain. It has been suggested such a drug could be useful for treating a wide range of conditions, from the withdrawal symptoms of drug addiction, to the management of pain or excessive stress.[25]

As we have seen, one critical difficulty with trying to create pharmacological intervention for obesity is the tremendous complexity of our brains and the interaction between hedonic and homeostatic systems. There is not one single area of the brain that must be targeted, but several. However, with the discovery of orexin, researchers may be getting closer to bridging certain gaps, and thereby to being able to treat the parts of the brain related to homeostatic and hedonic eating.

What the fMRI Study Revealed

Let's now return to the fMRI study we described at the start of this chapter. When Lauri Nummenmaa and his colleagues analysed the images taken of their subject's brains, they found that when an individual was overweight their brain responded differently to images of food.

The reward centres in the brains of both lean and obese women showed greater activation when looking at images of appetising food, as opposed to images of bland food. However, importantly, the level of this activation differed between the two groups.

'Responses to all foods (appetising and bland) were higher in obese than in normal-weight subjects,' explains Nummenmaa. 'When appetizing and bland foods were contrasted with each other, the caudate nucleus showed greater response in the obese subjects.' The results suggest that obese individuals' brains might constantly generate signals that promote eating even when the body does not require additional energy uptake.[26]

The response of the brains of obese people to food is characterised by an imbalance between those regions that promote reward-seeking and those involved in exercising self-control. Their brains are hypersensitive to the sight of food, which in a world replete with food cues is very bad news indeed!

In summary, obesity can be seen as the reflection of an imbalance between homeostatic signalling, reward signalling and inhibitory control within the brain. As people continue to put on weight, reward signalling becomes chemically disrupted, meaning that obese people start to have greater anticipatory response to food-related cues and imagery, which triggers the motivational cascade to engage in eating.

Lean individuals are able to limit eating in response to energy need; in essence, because their response to hypothalamic or homeostatic signalling is not overridden by acute sensitivity to food reward, they are able to ignore the external reminders prevalent in our environment that encourage overeating. For obese individuals, the pattern is not quite the same. There is a wealth of information that suggests obese people process food reward in a way that is distinct from lean individuals, and their biologically based heightened sensitivity to reward is likely the real culprit in disrupting eating motivation.

The overwhelming drive to pursue food reward, even at the risk of harming one's health, has been compared by some researchers to drug addiction.[27] Pinpointing why, precisely, some people are more vulnerable to the rewarding effects of food or other pleasurable activities has been the focal point in addiction-based research for decades. The first issue is that there is a heightened (hyper-) responsiveness to environmental cues that predict food; the second issue is that there may be hypo-responsiveness once food is actually consumed.

As we have seen, research has shown that, compared to lean individuals, overweight people show significantly greater anticipation prior to eating, yet it is not entirely clear whether obese individuals experience more pleasure while eating, or less.[28] We do know that obese individuals have a marked preference for high-fat and high-sugar foods and consume more of them.[29] Children with obese parents (who are therefore at a greater risk of obesity) also show a greater preference for the taste of high-fat, high-sugar foods and eat more avidly than children of lean parents.[30]

Several studies have reported that the obese show reduced levels of dopaminergic response to eating as compared to lean individuals.[31] These findings have led some researchers to suggest that overweight people become increasingly *hypo*-responsive to rewards and continue overeating to compensate for this depletion.[32] So the consumption of obese adults and children may in fact be tied to limited capacity to experience pleasure, which explains the desire to eat more in order to reach the 'normal' levels of pleasure enjoyed by lean people.

In our view, those *at risk* of becoming obese initially exhibit *hyper*-responsiveness in brain areas responsible for taste and touch. As a result, they anticipate their next meal more intensely and display greater sensitivity to food cues. However, as they start to put on weight, they develop a tolerance to the rewarding effects of food. This may be accompanied by a down regulation or 'pruning' of dopamine receptors in the basal ganglia. In response, the 'pleasure pedal' has then to be pushed down ever harder in order to achieve the same level of reward; in this case, more hyper-palatable food is needed to create the same sense of wellbeing and pleasure.

The idea follows similar models proposed for many other disorders of 'hedonic excess' – those disorders which relate to pleasurable substances or behaviours including gambling, shopping, sex addiction, or drug abuse. Yet obesity is unique, insofar as all systems in the body are implicated during the course of weight gain. It is not simply the brain, but also the body that contributes to the drive to overeat.

As we saw in Chapters 5 and 6, the obese person's greater resistance to leptin and insulin contribute to the sensation of hunger, yet it is the quest for pleasure which dopamine is responsible for that leads to a perceived need for ever-larger quantities of energy-dense foods.

Increased sensitivity to reward renders the obese individual more vulnerable to food cues, which are rife in our modern food environment. Paired with a numbed response to actual food ingestion, a perfect storm arises. Overeating and continued weight gain is caused by the combination of a predilection for energy-dense foods, an environment where such desires can be easily satisfied, and the body's inability to cope with this over-availability of fuel.

It is a cycle from which it is extremely difficult – though not impossible – to escape.

'As people become obese, they become insulin- and leptin-resistant, thus removing the normal peripheral signals that help inhibit the rewarding effects of food; the more severe the obesity, the worse the brain becomes at preventing excess food intake', explains Nora Volkow. 'You don't have a mechanism to counter the drive to eat . . . it's like driving a car without brakes.'[33]

CHAPTER 8

'I heard an ice-cream call my name': Why Impulsive Eating Means Overeating

'As I watch this kind of impulsive behaviour, I suspect a battle may be taking place in their heads, the struggle between "I want" and "I shouldn't," between "I'm in charge" and "I can't control this." In this struggle lies one of the most consequential battles we face to protect our health.'

David Kessler, *The End of Overeating*[1]

In the University of Sussex's Ingestive Behaviour laboratory, a young woman is inflating a total of thirty virtual balloons on a computer screen. With each click of the mouse button, a bright red balloon gets larger and she earns more money, and increases her probability of winning a $50 voucher. But each click is fraught with risk; if overinflated, the balloon will burst and all her earnings will be lost. Because each of the balloons has a different bursting point, she has no way of knowing the extent to which any one of them can be safely blown up. As we watch, she hesitates for a moment before risking a further click. The balloon swells but remains intact. Her look of relief rapidly changes to one of uncertainty. Should she click again in the hope of earning a higher reward, or stop and safeguard what she has earned?

Over the course of twenty days, sixty-four participants take part in this nail-biting game of dilemma, known as the Balloon

Analogue Risk Task or BART.[2] It is a challenge which allows psychologists to assess a participant's propensity for risk-taking and impulsivity, based on counting how many balloons they burst. The BART, a widely employed and well-validated test, shows clear differences in impulsivity between different categories of people.[3] It can, for example, discriminate between smokers and non-smokers, as smokers burst more balloons.[4]

Since the purpose of this research was to investigate the link between impulsivity and overeating, inflating balloons formed only part of the study. To their delight the participants also had to eat a chocolate ice-cream sundae and binge on a selection of delicious, high-calorie snacks. Why they were required to treat themselves in this way we will explain shortly.

Dietary Boundaries and Dietary Restraint

Although not widely known outside nutritional and dietary research, the concepts of dietary boundaries and dietary restraint have, for more than thirty years, exerted a powerful influence on psychological thinking about individual differences in eating behaviour.[5] A dietary boundary is a self-imposed limitation or rule for eating; it can be in the form of caloric restriction ('I must not consume more than 1,200 kcal'), or in the form of reduction of a macronutrient such as fat or carbohydrate.

When following a weight-control programme, dietary boundaries can be extensive and specific. For example, during the two-week-long induction stage of the Atkins Diet (the period when the greatest weight loss usually occurs), alcohol is banned, caffeine intake reduced to a minimum and carbohydrate consumption limited to fewer than 20 'net grams' per day. 'Net grams' means the total carbohydrate content of the foods, less the fibre – that number represents the grams of carbohydrate that have a significant impact on blood sugar levels. Foods with a good fibre-to-carbo-hydrate ratio are sometimes described as 'good carbs'.[6] Monitoring 'net carbs' and staying within defined limits of consumption is one kind of dietary boundary; others could refer to calories consumed, fat consumed, or any other imposed restriction to promote weight loss.

However, while theoretically a good idea, in practice these boundaries are also too easy to violate. Imagine that while following a diet, you are asked to dinner by a friend who happens to be an excellent cook. You are served a generous portion of homemade steak-and-kidney pudding, clocking in at over 400 calories. It is delicious and you clear your plate. Your host offers a second helping but your dietary boundaries include never having a second helping. However, to refuse might offend them. So, after a moment's hesitation, you take up the offer. Once again you finish every morsel. Dessert, a rich gateau served with ice cream, arrives. Normally you would politely decline, since another of your boundaries involves no high-calorie treats! On this occasion, aware of having already crossed your dietary boundaries you think, 'What the hell!', and accept two more helpings.[7] If you have ever struggled to control your weight, then this scenario will sound all too familiar.

The Restraint Scale is a tool used by psychological researchers to quantify self-control in the realm of eating behaviour. Individuals who set themselves a large number of dietary boundaries score 'high' on this scale. While being able to control what you eat should obviously be a good thing, researchers have speculated that such people may try to consume fewer calories than their body actually needs for healthy metabolism, leading to feelings of deprivation. Therefore, the self-imposed boundaries can, in some cases, lead to an increased risk of binge eating and development of eating disorders such as bulimia.[8] During the 1980s, in order to gain a greater understanding of dietary restraint, Peter Herman and Janet Polivy from the University of Toronto gave identical bowls of ice cream to both restrained and unrestrained eaters. They found, to no one's great surprise, that the subjects who had been rated as more restrained subjects ate significantly less than the latter. Yet when they instructed both groups to drink a calorie-rich milkshake immediately before eating the ice cream, the outcome was unexpectedly reversed. In that scenario, restrained eaters consumed twice as much as unrestrained ones. This brings us back to the surprising findings of our balloon-bursting study, and the importance that a person's psychological profile has on eating behaviour.

BART and Impulsive Eating

Before being selected for the study, all the volunteers – none of whom were overweight – completed a questionnaire designed to provide information about their attitudes towards food in general and dieting in particular. They were asked, for example, to tick True or False to statements such as:

- While on a diet, if I eat food that is not allowed, I often then splurge and eat other high-calorie foods   True/False
- If I eat a little bit more one day, I make up for it the next day   True/False
- I eat diet foods, even if they do not taste very good True/False
- A diet would be too boring a way for me to lose weight True/False

Their responses placed them in one of two groups. Group A consisted of women who had tried, often on several occasions, to go on a formal diet (Unsuccessful Dieters), or whose weight tended to yo-yo (Weight Fluctuators). These two subgroups together were comparable to 'What-the-Hell' type dieters who, after encountering a challenge with dieting, have a tendency to throw in the towel. In Group B were those who had either succeeded in dieting (Successful Dieters) or were able to maintain a healthy weight while imposing few, if any, restrictions on what they ate (Unrestrained Eaters). We referred to them collectively as 'Successful Dieters'.

We compared the reactions of the What-the-Hell types, to the Successful Dieters to investigate whether their eating habits could be linked with other kinds of traits, such as impulsivity.

The participants in the study had been asked not to eat or drink anything after 11 p.m. the previous night, which meant they were all distinctly hungry on arriving at the lab. There they were given a light breakfast, comprising some orange juice and a bowl of Crunchy Nut Cornflakes with semi-skimmed milk amounting, in total, to exactly 400 calories.

After breakfast, each was weighed. They were then asked to refrain from eating and to drink only water before returning, three hours later, to undertake a variety of tests and to consume a chocolate and ice-cream sundae. Apart from ensuring the volunteers felt comfortably full, this 370-calorie snack had a more devious purpose. As in Peter Herman and Janet Polivy's study, we were compelling the subjects to violate their dietary boundaries, by providing them with high-calorie treat and insisting they ate the lot. After they had finished their chocolate ice-cream sundae, the volunteers were offered a selection of high energy-dense snacks, including chocolate buttons, biscuits and dry-roasted peanuts. They were invited to consume as many or as few as they wished.

As predicted, the What-the-Hell Dieters ate upwards of 300 kcal more snacks than Successful Dieters and Unrestrained Eaters. Following this snacking binge, the volunteers then undertook the virtual balloons test once more.

It had been expected that after eating so much high-fat, high-sugar food – some participants had consumed upwards of 800 calories of pure sugar – both groups would become more impulsive. This was based on the assumption that What-the-Hell Dieters failed to stick to diets because they ate impulsively anyway, and that because they had already been compelled to violate their dietary boundaries, Successful Dieters would feel as though they could continue to do so.

This was not, however, what actually happened. While, as predicted, the What-the-Hell group behaved *more* impulsively prior to eating HED foods they became *less* impulsive after doing so! It was the Successful Dieters who became more impulsive after the high calorie snack, who took greater risks and so burst more balloons.

So eating energy-dense foods actually made people who usually have difficulty controlling their impulses better able to do so. Might this suggest that naturally impulsive people sometimes eat to excess because this makes it easier for them to exert self-control in other situations?

While the fact that people 'comfort-eat' when upset or stressed is well known (see Chapter 10 for a more detailed discussion), the notion that overeating might be used to reduce impulsivity had not previously been investigated. These results suggest something interesting and important: people who have great difficulty controlling their weight are not necessarily hungrier; rather their psychological profile and life circumstances might be the key contributing factors that lead to overeating, as they may be using food as a psychological crutch.

So, if you see yourself as a weight cycler, it's very likely you may unconsciously use food to help you concentrate or to overcome stressful situations. You may have a tendency to snack at your desk during times of stress, or you may find you are less irritable when junk food is available.

The way to reduce the gnawing cravings is to simply not allow yourself to give in. That, of course, is much easier said than done. The key is to try to alter other elements of your lifestyle to give yourself a greater chance of resisting your impulsive urges. By getting more sleep and incorporating exercise into your daily routine, you can take significant steps in this direction.[9]

For example, in an investigation of over 2,000 New England homes, it was found that moderately active to very active people tended to consume more fibre, less total fat, higher dietary fibre, reduced cholesterol, and follow diet guidelines more carefully than less active people.[10] Yet it's not simply the fact that healthy people engage in healthy behaviours, rather that scientists are beginning to suggest that physical activity may reduce impulsivity via neuro-cognitive mechanisms, which would allow a person to avoid eating impulsively without having to register the act mentally. They would just do it.

Decision-making, regardless of what the task at hand may be (whether eating or otherwise) is evaluated rationally, or influenced by emotions, and a surge of data demonstrates that physical activity improves mood. Improved mood may yield fewer impulsively based decisions for eating. Moreover, exercise is positively corre-lated with neurogenerative, neuroadaptive, and neuroprotective process.[11]

For example, a study of adults who completed a six-month walking regime showed they not only became more aerobically fit, but showed improved decision-making. This cognitive improve-ment was defined by their performance on selective attention tasks, and 'stop signal' paradigms which measure an individual's propen-sity for impulse control.[12]

If exercise helps reduce impulsivity, it seems a safe assumption that a lack of physical activity will lead to greater impulsivity. This is indeed the case. In a unique investigation at Berlin's Centre for Space Medicine, scientists investigated twenty-four males who were put in a bed-rest model of 'prolonged weightlessness', with the bed slightly tilted at 6 degrees. Their scores on a psychological tool to assess impulsivity and risk-taking, known as the Iowa Gambling Task, showed greater impulsivity after the session of weightlessness than before. In a real-world context, this may explain the link between sedentary behaviour such as TV viewing and increased eating.

To summarise, physical activity can help build cognitive resources, such as inhibitory control, which can ultimately contribute to tighter control of eating motivation, and eventually a reduced need to eat for emotional reasons.

The Overeating Impulse

Impulsivity was once regarded as a fixed personality trait, something you were born with, like blue eyes or blonde hair, but we now know it depends as much on context as on inborn characteristics – individuals able to exercise great self-restraint in one situation may find it almost impossible to do so in another.

A large and fast-growing volume of research has also demonstrated the extent to which impulsivity significantly increases an individual's risk both of becoming obese and developing a range of eating disorders, including both binge and binge/purge eating.[13] Impulsivity is not only a precursor to overeating, but may also be linked to a person's perception of, and sensitivity to, the rewarding properties of HED foods as discussed in the previous chapter.[14] Foods with high levels of sugars and fats can trigger a craving to consume them again – as dopamine levels dissipate, the person is left wanting more.[15] The fact we use the word 'craving' to describe an intense desire in the context of both food and illegal drugs is no mere accident of language; as we have seen, the impulse to overeat is produced in the same brain regions as those associated with drug addiction.

Why Self-Control Fails

Why do some people find it so hard to exercise control over their impulses when it comes to eating healthily or following a diet? The main factor, according to some psychologists, actually lies in the process of self-control itself.

'Self-control resembles a muscle in more ways than one,' believes Roy Baumeister, Professor of Psychology at Florida State University in Tallahassee. 'Not only does it show fatigue, in the sense that it seems to lose power right after being used, it also gets stronger through exercise.'[16]

In a 1998 study conducted by Roy Baumeister and his colleagues from Case Western Reserve University, sixty-seven psychology students (thirty-one men and thirty-six women) were recruited to participate in what they believed was a taste perception test. Arriving at the laboratory they were shown two types of food. One was a pile of freshly baked chocolate chip cookies, whose mouthwatering aroma filled the laboratory, the other was a bowl of red and white radishes. They were told that these had been chosen for the study because they were 'highly distinctive foods, familiar to most people.'

The students were then assigned to one of four groups. The first was asked to taste the chocolate chip cookies, and a few radishes. The second to eat only the cookies, the third only the radishes. Those in the final group were offered no food at all and moved directly to the second part of the study, which involved problem solving. Having issued these instructions, the psychologist left the room but continued to observe the groups through a one-way mirror, recording the amount of food eaten and checking that the three groups ate only the foods assigned to them. After five minutes she returned to set all four groups the task of solving a problem, which was in fact unsolvable. It involved trying to trace a geometric figure without going back over any of the lines or lifting the pen from the paper. They were told: 'You can take as much time and as many trials as you want . . . If you wish to stop before you finish [i.e. before you've solved the problem], ring the bell on the table.'[17]

This, although the students did not know it of course, was the crux of the experiment. The hypothesis that Baumeister and his team was testing was that those compelled to eat only radishes and watch while their more fortunate companions enjoyed the cookies would give up on the puzzle test faster. In short, they thought having to exercise self-control to ignore the tempting cookies would then leave them less able to struggle against their frustration with the impossible task. The psychologists noted that while none of those in the 'radish only' group grabbed a cookie when left alone in the room, several, in Baumeister's words, looked 'longingly at the chocolate display and in a few cases even picked up the cookies to sniff them'. Subsequently these 'deprived' students spontaneously admitted they had experienced great difficulty in resisting the temptation to eat the cookies.

The results bore out the researcher's hypothesis. Those who had been allowed to eat cookies made 34 attempts to solve the problem over a period of nearly 19 minutes. Those who had not been offered any snacks at all made only slightly fewer attempts (32) over a somewhat longer period (21 minutes). However, the radish only group did, as predicted, abandon the challenge far earlier. They made only 19 attempts and gave up after just 8 minutes.

'Resisting temptation seems to have produced a psychic cost,' notes Roy Baumeister, 'in the sense that afterward participants were more inclined to give up easily in the face of frustration. It was not that eating chocolate improved performance. Rather, wanting chocolate but eating radishes instead, especially under circumstances in which it would seemingly be easy and safe to snitch some chocolates, seems to have consumed some resource and therefore left people less able to persist at the puzzles.'[18]

Variations on this type of experiment have consistently supported the finding that exercising restraint in one situation reduces one's ability to exercise it in another. In a more recent study, participants whose self-control had been depleted chose relatively trashy films over more intellectual or artistic movies. These preferences occurred even when selecting a film to watch in several days' time.[19] And in another study, participants whose self-control had been depleted favoured sweets over healthier granola bars as a snack. The amount of food consumed under these circumstances was also affected. Dieters ate more food when their self-control had been depleted than they would otherwise have done, while those not on a diet remained relatively unaffected.

'The distinction is important,' points out Roy Baumeister, 'because it suggests that . . . depletion does not simply increase appetites or pleasure seeking. Rather, it undermines the defences and the virtuous intentions that would otherwise guide behaviour.'[20]

The previous study demonstrates a key point: if you employ a lot of self-control in one area of life then you will find it harder to do so in other areas. So someone who struggles with impulsive overeating will find it harder to resist those impulses if they have already depleted their capacity for self-control by using it in other contexts. When we begin to understand self-control in this way, as being almost like a finite resource, then we see that in some situations a person can no more prevent themselves from overeating than a car can keep running when the petrol tank is empty.

Self-control, however, does more than keep our impulses in check. It forms part of a much larger collection of executive functions concerned with self-monitoring, coping with stress, considering different options, weighing up alternatives, and making decisions. And all of these draw on the same limited 'resource' of our character as well. It may be the case that certain times are fraught with so much stress that it limits our abilities for self-control. Thus, learning to exercise self-control and expand it like a muscle is very likely to contribute to a better relationship with food, even in difficult times.

Self-control depletion can also come about merely by watching others exert willpower. In a recent study by Joshua M. Ackerman of Yale University and his colleagues, undergraduates were asked to read a story about a hungry waiter or waitress (matched to the sex of the participant) working in a high-quality restaurant, who was forbidden to eat while at work on pain of being fired. The story, written in the first person, described in great detail all the mouthwatering dishes being served and how difficult it was to resist the temptation of sneaking a morsel to assuage their appetite. Half the students were told to read the story while the other half were asked to try and 'really imagine yourself in his or her shoes, and concentrate on trying to imagine what the person was thinking and how he or she was feeling.'[21] They were then shown pictures of various mid- to high-priced products like watches or cars, and asked how much they would be prepared to pay for them. The results were surprising. Those who had been asked vividly to imagine the plight of the waiter, and truly immerse themselves in the waiter's situation, were willing to spend more than $6,000 more than participants who had simply read the story.

'The findings from the current research suggest that the ability to control one's own thoughts, feelings, and behaviours is influenced by the self-control of other people,' wrote Joshua M. Ackerman, 'and by how closely one's mind mirrors the minds of others, in ways one might not generally expect.'[22] This, and similar studies, offers a caution to anyone who believes that joining a group of fellow dieters will make it easier to resist temptation – striving to exercise self-control while watching others do the same may, in fact, make it harder.

'Self-control is one of the defining features of the human animal,' according to Michael Inzlicht of the University of Toronto and Brandon Schmeichel of Texas A&M University. 'Its failure is one of the central problems of human society, being implicated in phenomena ranging from criminality to obesity, from personal debt to drug abuse.'[23]

So if it is so important to our health and wellbeing, why does self-control so often fail? Often, we have to look at peripheral factors, such as sleep loss, that may contribute to our inability to withhold from immediate desires.

Sleep Loss and Weight Gain

Over the past ten years, the number of hours people spend asleep has significantly declined across many parts of the world, just as obesity levels have dramatically increased.[24]

Could these two events be related?

Epidemiological studies propose a U-shaped relationship between the two, meaning that both having too little or too much sleep leads to an increase in body weight. Sebastian Schmid and his colleagues from the University of Lübeck and the Interdisciplinary Obesity Centre in Rorschach in Switzerland, hypothesised that even one night's disturbed sleep would produce a rise in levels of ghrelin, the hormone whose role in appetite regulation we looked at in Chapter 6.

To put their theory to the test, they separated ten normal-weight young men into two groups. One group they allowed to sleep undisturbed for seven hours and the other for just four and a half hours. Would this small amount of sleep deprivation change ghrelin levels? Blood tests showed that it did. Sleeping for only four and a half hours on a single occasion produced increases in feelings of hunger and in ghrelin levels in the blood.[25]

We conducted a study of our own in this area.[26] In it, we also explored the extent to which sleep deprivation would affect a group of normal-weight young men and women. Our volunteers were invited to a hotel just outside London to take part in what they believed would be a series of team-building exercises. Because an early start was needed, they were checked in the night before and were divided into a Blue Team and a Yellow Team. At one a.m. a member of our research team hammered on the bedroom doors of the Yellow Team and told the bleary-eyed occupants to tackle a tough IQ test. This took about 30 minutes to complete, after which they resumed their disturbed sleep. But not for long – two hours later they were again roused from their slumbers, to complete a second IQ test. Once they had finished this they went back to bed and slept until 7.30 a.m., when both groups had to get up and come down to breakfast.

Throughout the night, using infrared cameras, we had watched the 'sleep-disturbed' participants to see just how effectively their normal sleep patterns were being disrupted. Taking into account the time they had spent struggling with the IQ tests, we calculated that, on average, the eight hours of sleep enjoyed by the Blue Team had been slashed to five for members of the Yellow Team.

After breakfast the following morning, both the well-rested Blue Team and the weary Yellow Team were handed bags filled with snacks on which to nibble throughout the day. Some were high energy-dense snacks, such as chocolates and biscuits, as well as healthier options including apples and low-calorie fruit bars. For the rest of the morning the teams then competed against one another in various games, which they believed to be the purpose of the study. What we were actually interested in was, of course, whether those deprived of sleep would impulsively consume more of the snacks high in sugar and fat than their well-rested rivals.

They did. The sleep-deprived Yellow Team consumed a third more calories than members of the Blue Team; losing sleep on even a single night can result in more impulsive eating the following day. Chronic sleep deprivation, therefore, will place an even greater burden on self-control, making weight gain more likely and obesity a stronger probability.

The dangers of not getting enough sleep in relation to obesity have been recognised by others. 'Insufficient sleep (short sleep duration and/or poor sleep quality) has become pervasive in modern societies with 24/7 availability of commodities,' explains sleep and obesity expert Jean-Philippe Chaput from the Children's Hospital of Eastern Ontario. 'Factors responsible for this secular decline in sleep duration are numerous and generally ascribed to the modern way of living (e.g. artificial light, caffeine use, late-night screen time, parental attitudes).'[27]

Dr Chaput highlights six ways in which lack of sleep contributes to future bouts of overeating. These are:

- More time and opportunities for eating
- Psychological distress
- Greater sensitivity to food reward
- Disinhibited overeating
- More energy needed to sustain wakefulness
- Changes in appetite hormones.

Children, too, can become more impulsive when their sleep is disturbed, especially if this is due to what is known as Sleep Related Respiratory Disorder (SRRD), which can cause them to snore or have difficulty breathing when asleep. A third of school-age children (35%) are reported to suffer from disordered sleep of whom three out of a hundred have SRRD and up to one in six (16%) are impulsive.[28]

To investigate the relationship between impulsivity and sleep disorders among children, Marilaine Medeiros and her colleagues from the Sleep Laboratory at São Paulo hospital studied 1,180 children, 547 of whom had poor sleep while the remainder slept normally.

'School-age children with impulsive and dysfunctional behaviours show difficulties in falling asleep, refuse to lie down, present feelings of anxiety or fear at bedtime, stop taking a nap at an early age, and show nocturnal agitation, frequent nocturnal arousals and/or difficulties in falling asleep again,' she reports. The aim of her study was to determine, 'whether impulsivity is more prevalent in children with sleep disorders than in those without sleep disorders, and within which kind of sleep disorder.'[29]

The results were as Medeiros expected: children who slept poorly were also more likely to display impulsive behaviour, especially when their disturbed rest was due to impaired breathing. Consider the implications of this. If one of the ways that a child's impulsiveness expressed itself were to be by overeating, then they could easily become trapped in a vicious circle; being overweight makes it harder to breathe efficiently, making it harder to sleep properly. Impulsive behaviour due to fatigue would then lead to further overeating, causing further weight gain and therefore making it even harder to breathe and sleep properly – and so on. If such a cycle starts in childhood, the probability is high that it will continue beyond adolescence and into adulthood.

Obesity is strongly linked with a variety of respiratory symptoms and diseases, including obstructive sleep apnoea syndrome (in which the individual momentarily stops breathing); obesity hypoventilation syndrome, asthma and decreases in lung volumes.[30] The increased weight on the chest wall resulting from being overweight makes it harder to breathe in a relaxed and easy manner. These difficulties are compounded by greater resistance in the airways and a build-up of fat around the abdomen and intra-abdominal tissue, which impedes movement of the diaphragm.[31]

Impulsivity and Overeating

While impulsive eating is sometimes defined as 'eating which takes place outside of regular meals or snacks', as Michael Lowe of Drexel University and Kathleen Eldredge of Stanford University School of Medicine pointed out in 1993, many people's lives include instances of this kind of eating without it becoming problematic.[32]

In fact, occasional impulsive snacks should not pose a direct threat to maintaining a healthy body weight. As we've already mentioned, the problems occur when we begin to use food to diminish other impulsive tendencies. Since we are constantly in environments that require self-control, including school, work, and social situations, it is increasingly important for us to identify situations that trigger the 'need' to eat energy-dense, nutrient-poor foods.

Yet, as we have shown, our food environment itself can contribute to this problem. The mere mention of a fast-food chain elicits impulsive behaviour, and our inhibitory control is tested virtually every time we walk into the supermarket – we are having to perpetually say 'no' to ourselves in the face of a mounting concentration of hyper-palatable foods. Whether in the grocery aisle, fast-food queue, or at a table in a restaurant, we are now required to put the brakes on our drive to eat with tremendous frequency. With such demands on our self-control coming so thick and fast, it's easy to see how factors which erode it, such as insufficient sleep, can have tremendous repercussions.

Only by focusing on impulsivity as a key factor in the obesity epidemic can we begin to understand how to avoid, or perhaps control, behaviours that contribute to obesity.[33] By improving sleep and exercise habits, and minimising stress, it is possible to increase reserves of self-control – which, as we have seen, is a finite psychological resource. Doing this, and also striving to identify and avoid situations which make us react in an impulsive way, will greatly increase anyone's ability to control overeating.

Admittedly, 'bolster your cognitive resources' is hardly a match for more easily recalled public health slogans of the past such as, 'eat less, move more', or the 'five a day' plan. But it really does seem that we should be making cognition and impulse control a primary focus in the battle against obesity. When we begin to look at things from this angle we also gain a greater appreciation of the complexity of the issue. Obesity can lead to fundamental changes with the way a person perceives, experiences and desires food. There is, as shown in Chapter 7, a clear difference between obese and lean people in their responses to food-related imagery, with the former experiencing heightened anticipation and reduced impulse control.

The differences between lean and obese reactions to food-related imagery are so profound that some researchers have suggested that obesity could be considered a brain disease.[34] But whether obesity is the result of brain impairment in the first place, or of over-nutrition contributing to subsequent cognitive problems, isn't really the point. There is growing evidence to suggest that over-eating can contribute to inflammation within the brain.[35] In animal studies it has been shown that over time a high-fat diet leads to inflammation of the brain's outer layer, the cortex, and of the hypothalamus.[36] Such a diet also contributes to impaired insulin secretion and sensitivity.[37] All of these disrupt an individual's ability to guide their eating to match the needs of their body – rather they are bound to satisfying a deep-rooted desire for pleasurable experience. Thus, eating impulsively is the net result of significant biological changes that will perpetuate the cycle of overeating.

The truth is that we are not all on a level playing field when it comes to controlling our eating. The 'greed' which the obese are often accused of is a far more complex problem than that inaccurate and stigmatising label suggests.

CHAPTER 9

Feeding Our Feelings:
The Power of Emotional Eating

'I don't stop eating when I'm full. The meal isn't over when I'm
full. It's over when I hate myself.'
Louis C.K., comedian[1]

The cold, grey, November afternoon had done nothing to dampen
the excitement of the Bath and Exeter rugby supporters as their
teams competed in what turned out to be a fiercely contested
end-of-season match. At the final whistle, Bath had trounced Exeter
by 37 points to 15, leaving Bath fans jubilant and Exeter supporters
downcast.

Just the sort of powerful emotions we had been hoping for!

We were about to carry out an experiment into the link between
emotional arousal and overeating.[2] In this experiment, groups of
fans from each team were invited to attend a post-match analysis,
hosted by well-known rugby pundits, in a nearby hotel. On arrival
they were taken to separate meeting rooms and offered a buffet
of pizzas, crisps, sausages and chicken nuggets, as well as bowls
of salad and fruit. They could eat as much as they wanted and
return to the buffet as often as they wished.

The discussions they had while eating were lively, informed and,
at times, heated. What the fans did not know was that we were
more interested in what went into their mouths than what came
out of them. Previous research had shown that sports fans in
America and France ate significantly more high fat and high sugar

foods when their teams lost.[3] The question we wanted to answer was, would English fans do the same?

They certainly did. Between them, supporters of the victorious Bath team consumed 8,000 calories, while defeated Exeter team fans ate almost 60% more at 13,412 calories.

But why?

What is the link between one's emotional state and a desire to binge on highly palatable, energy-dense but nutrient-poor food? 'It's a form of self-medication', according to George Koob of The Scripps Research Institute in La Jolla, California. 'You're modulating your arousal. People take the food to calm themselves down . . . In other words they relieve the itch.'[4]

In order to understand how this works, we need to look at the two very different reasons why we eat.

Homeostatic vs. Hedonic Eating

We've already touched on the ideas of homeostatic and hedonic eating in Chapter 7. The former, as the name suggests, is when we eat in order to maintain 'homeostasis', the essential internal balance on which survival depends. Homeostatic eating is when we eat in order to reach a genetically determined metabolic 'set point' and, as discussed earlier, it is primarily guided by the hypothalamus.

Hedonic eating, on the other hand, involves feeding not our physical appetite and the needs of our body, but rather our desires – the word 'hedonic' is derived from the ancient Greek *hedone*, meaning 'pleasure'. A lot of hedonic eating takes place when we are trying to cope with either negative or positive emotions. We may, for example, use chocolates, cakes, biscuits or doughnuts as comfort foods when feeling bored, rejected, frustrated or depressed. Researchers have found, for example, that obese people consume significantly more chocolate when watching a sad film than a neutral one.[5] However, we also eat hedonically when excited or as an aid to relaxation – take the super-sized bucket of popcorn that adds to our enjoyment of a trip to the cinema.

The foods we prefer when eating hedonically are almost invariably highly palatable and energy dense. And if we are eating for emotional reasons as opposed to physiological ones, we are rendering ourselves vulnerable to the excessive amounts of food available to us in the modern world; we are being tempted with the offer of the fix we crave almost constantly by advertising and displays of tasty treats. Further, the more prone a person is to emotional eating, the more sensitive they are to these food-related cues. Hypersensitivity to food cues, paired with an environment of plenty, creates the perfect storm, where weight gain is an obvious and – without appropriate behaviour modification tools – inevitable outcome.

Hedonic Eating Is Often Mindless Eating

In an ideal scenario, once we have eaten enough to satisfy our homeostatic needs, signals from the stomach will alert the brain to the fact that sufficient food has been eaten and those signals are attended to. Most people know when they have eaten enough and stop before feeling uncomfortably full. Furthermore, since the food is satisfying a physical need, the individual has no reason to feel regretful or ashamed.

This is not necessarily the case with hedonic eating; since food consumed for emotional reasons is usually hugely energy dense, our stomachs do not react to the energy content appropriately.[6] If we compare 50 g of vegetables to 50 g of chocolate, we'll notice the chocolate takes up significantly less space. Chocolate is immensely energy dense, which makes it difficult for the body to know when 'enough' has been ingested. This is typically the case with most junk foods; they are amazingly efficient vehicles of energy delivery, which means it can be easy to eat too much and thus gain weight.

Someone feeding their feelings might easily finish off an entire jumbo-sized bag of crisps because their appetite is not guided by physical sensations of satiety so much as a gnawing psychological need. Moreover, hedonically motivated eating can elicit feelings of guilt and shame; people aren't sure *why* they have engaged in a binge, but do it in spite of their best intentions.

In her book *Counting Calories*, author Jane Olson discusses her own struggles with hedonic eating: 'Comfort foods they may have been, but helpful foods they most definitely were not. By merging my identity with certain foods and thinking of them as old friends, I found myself in the food equivalent of a co-dependent, destructive relationship. If I was going to insist on relating to food as a friend, then clearly I needed new friends.'[7]

The Pleasure Triggers for Hedonic Eating

In a 2003 survey conducted by the American food magazine *Bon Appétit*, respondents ranked their favourite foods in order of preference as: ice cream, chocolate, cake, cheesecake, and potato chips.[8] If a similar survey were conducted today in the UK or mainland Europe, our guess is that a similar order would be reported. In order to understand what makes these foods so special and so pleasurable when it comes to hedonic eating, we need to count up the number of 'taste triggers' (a term used in the food industry) that each possesses. There are six main taste triggers and it will probably come as little surprise to you when we say that food companies carefully design these into certain products in order to create what they call 'bliss' points – combinations of taste triggers that can make the foods so delicious they are arguably addictive.

Trigger One – Flavour: For maximum hedonic pleasure, food should contain salt, sugar, monosodium glutamate (MSG), and flavour-active compounds. Surprisingly, taste only accounts for 10% of the overall *sensation* of food. The optimal level for salt is between 1.0 and 1.5%; for MSG 0.15% and for flavour-active compounds 0.02%.[9]

Trigger Two – Dynamic Contrast: Many top-selling hedonic treats have high texture and flavour contrasts. Chocolate and ice cream, for example, start as a solid then melt into a liquid in the mouth; jam doughnuts combine fluffy dough with the satisfying wetness of the jam at the centre. The greater the contrast, the more pleasurable and rewarding the food will prove.[10]

Trigger Three – Evoked Qualities: The ideal food should evoke agreeable memories about where it was first enjoyed and our physiological and psychological state at that time. For example, the Valentine's Day chocolates given by an admirer, the ice cream enjoyed on holiday, the scent of birthday cake candles blown out as a child. Even when no such actual memories exist, they can often be created through advertising and marketing.

Trigger Four – Food Pleasure Equation: The food pleasure equation is a calculation made in the food industry involving the relationship between the sensation of a food and the macronutrient stimulation it causes, the latter being to do with the combination of carbohydrate, fat and protein. The nicest foods maximize both elements, and if one is reduced in some way it will have to be compensated for in order for the food to remain as tasty – so if manufacturers reduce the fat in a product they have to enhance its palatability in some other way to ensure the same level of hedonic appeal. Typically this is achieved by increasing amounts of sugar and/or artificial sweeteners.

Trigger Five – Caloric Density (CD): Is a measure of how many calories there are in a food relative to its weight. On a scale in which water scores 0 and pure fat 9, food with a CD of between 4 and 5 is the most palatable.[11] As we explained in Chapter 7, our gut has a variety of receptors, and foods with this ideal CD stimulate these to generate feelings of intense pleasure in the brain.

Trigger Six – Emulsions: The word, which comes from the Latin 'to milk', is used to describe a mixture of two or more normally immiscible (i.e. non-mixable or unblendable) liquids, such as the water and fat in milk itself. Examples of emulsions in food include ice cream, chocolate, butter, vinaigrettes, mayonnaise, and brewed coffee. Emulsions raise the pleasurable sensation of high fat/high sweet foods to a new level. For example, while butter contains about 2.5% salt in the solid state, the actual salt content experienced by the taste buds jumps to about 10% in the mouth.

Chocolate Heaven – Chocolate Hell

One food which possesses all these pleasure triggers is chocolate.[12]

It was Christopher Columbus who, in the early years of the sixteenth century, brought cocoa beans to Europe from the New World. But it was the Spanish conquistador Don Hernán Cortés who first realised their huge commercial value. When initially introduced to Europeans, chocolate was marketed as an enjoyable drink with medicinal properties; it was said to cure a variety of health problems ranging from stomach ache to depression. Today, although they may not realise it, many of those who comfort-eat chocolate are using it for the same reasons.

Chocolate manufacture and sales is a multibillion-dollar

business, the value of which exceeds the Gross Domestic Product (GDP) of more than 130 nations. Americans consume over $13.1 billion in sales annually, amounting to three billion pounds of chocolate.[13] The annual per capita consumption of chocolate in the States is 13 lb (6 kg) and more than half (52%) of adults prefer the flavour of chocolate to both vanilla and fruit.[14] Europe loves chocolate even more, with an annual per capita consumption a shocking 22 lb (10 kg). At the other end of the scale, India and China's chocolate consumption rates are among the lowest; the Chinese eat less than 6 ounces a year and the Indians just 3.5 ounces. These low numbers are, however, likely to rise sharply in coming years as the economies of these countries continue to grow, and their populations' taste for indulgent treats along with them.[15] So let's take a look beneath the wrapper to see what it is about chocolate's chemistry that makes it both powerfully appealing and seriously addictive to men and women alike.

Few of the millions of chocoholics around the world appreciate the potent effect their favourite treat has on their brain. Chocolate has a psychoactive effect so powerful some researchers believe it should be considered not confectionary but a drug. Among other compounds, a piece of chocolate contains:

- Caffeine: A psychoactive substance that produces temporary alertness.
- Tryptophan, an amino acid and serotonin, a hormone, which in concert generate a sense of wellbeing and relaxation.
- Xanthines: A mild stimulant which occurs naturally in the brain and which, like caffeine, increases alertness.
- Theobromine: A stimulant which increases blood flow.
- Phenylethylamine: Sometimes dubbed a 'love chemical', this compound stimulates the brain to release dopamine which, as we explained in Chapter 8, produces feelings of euphoria and

an intense desire to repeat the experience as quickly as possible.

- Flavonols: compounds which boost the flow of blood to the brain for up to three hours after being consumed. Also found in red wine, blueberries and green tea, these aspirin-like compounds also exert a mild analgesic effect.
- Chocolate fat contains chemicals similar to a compound in the brain called anandamide, a neurochemical that binds to marijuana Cannabinoid1 (CB1) receptors and produces a natural and legal high.[16]
- Chocolate and cocoa also contain interesting psychoactive compounds called tetrahydro-beta-carbolines. These are present in all types of chocolate, and also, in appreciable quantities, in cereals containing chocolate. The darker the chocolate, the higher the tetrahydro-beta-carboline content.[17] They belong to a class of substances known as alkaloids, which include psilocybin (present in magic mushrooms) and reserpine. Both have been used in religious ceremonies and medicine for thousands of years. Tetrahydro-beta-carbolines mildly inhibit the production of an enzyme called monoamine oxidase (MAO),* which prevents the breakdown of monoamine receptors, making dopamine, norepinephrine, and serotonin more available. By inhibiting MAO activity, chocolate increases the levels of these mood-enhancing substances, making people feel more cheerful and optimistic. Its effect is the same, although milder, than that of SSRI antidepressants (Selective Serotonin Reuptake Inhibitors) such as paroxetine, which are used to treat a wide range of disorders including depression, anxiety, and post-traumatic stress.

---

* Monoamine Oxidase inhibitors (MAOI) are a group of medicines used to treat depression. They include: isocarboxazid, phenelzine, moclobemide and tranylcypromine and appear under various different brand names.

In 2006, Gordon Parker and Joanna Crawford from the Prince of Wales Hospital in Sydney, Australia, reviewed all the possible theories as to why people develop chocolate cravings.[18] After analysing the responses of almost 3,000 adults with clinical depression, they concluded that the fat and sugar activate both dopamine and opioids (the latter being the term for any chemical which has similar effects to morphine and its derivatives) in a strong but dysfunctional manner and that this results in overconsumption.[19] Which is to say that people who do suffer with depression may be vulnerable to overeating chocolate because it enhances mood. This dopamine-opioid rush, in combination with calories in the form of fat and sugar, also has effects on receptors in the gut. This is why, like any addictive drug, eating chocolate even once increases the likelihood of it being eaten again and again.

Rewards and the Obese Brain

Research into addictive behaviour indicates that dopamine receptors in the striatum (in a part of the brain called the basal ganglia, see Chapter 7) work in tandem with frontal regions in the brain, where decisions about whether or not to eat are made. Obese individuals show an enhanced response in this region of the brain when anticipating food, which increases their risk of overeating.[20]

Furthermore, dopamine receptors will eventually shut down if they are overstimulated. As with drug addiction, repeated over-consumption of highly palatable foods can result in long-term neuroadaptations in the brain's reward and stress pathways. The brain seeks to maintain homeostasis, normal balanced operations, by removing or blunting the sensitivity of dopamine receptors. This means that larger and larger quantities of the rewarding substance are required to produce the same pleasurable effect. This is why a junkie has to constantly increase doses of heroin or cocaine to experience the same high. In the same way, a person who is gaining weight from bingeing on chocolate or other HED foods will find that they too must increase the amounts they consume in order to get the same satisfaction. This, inevitably, diminishes their ability to control the amount they eat and significantly increases their risk of substantially overeating.[21]

Binge Eating – When Food Becomes Addictive

In a study of sugar bingeing, Nicole Avena, Professor of Psychology at Columbia University, placed rats on a daily schedule of 12 hours without food, followed by 12 hours unlimited access to a sugar-rich diet consisting of standard rodent chow plus a solution of 25% glucose or 10% sucrose.[22] After only a few hours, they started to binge on the super-sweet solution, their desire for sugar having apparently surged. They also showed an increased desire to obtain sucrose, pressing the lever which delivered the treat more vigorously after being deprived of the extra sugar for a couple of weeks.[23] The rats' daily feeding patterns also shifted as they started to consume more and more sugar meals.

'After a month of binge eating sugar,' Avena reported, 'rats show a series of behaviours similar to the effects of drugs of abuse, including the escalation of daily sugar intake and increase in sugar intake during the first hour of daily access.'[24]

The similarities with drug addiction was further demonstrated when the researchers injected rats that had been binge eating sugar with naloxone. This opioid-receptor antagonist is used on humans to counter the effects of heroin or morphine. After receiving the drug, the rats showed withdrawal symptoms, such as teeth chattering, forepaw tremor, and head shakes.[25] As with any other addicts, the animals had become so hooked on sugar they had to have more and more of it in order to achieve the same level of reward and satisfaction. So sugar is not only a source of energy, but a potentially addictive source of chemically based pleasure.

One of the most important factors in the development of binge eating is a prior history of food restriction or energy deprivation. Energy deprivation can serve to enhance the sensory qualities of food even further, meaning that by going on a diet a person can inadvertently make themselves more likely to indulge in binge eating if their self-control snaps. Moreover, the depression or negative mood experienced when food intake is limited lifts immediately when we eat foods that are high in sugar and fat. This improvement is likely due in part to a surge in dopamine production.[26]

This supports the notion that, during a binge, emotion-driven eaters are especially powerfully attracted to what they regard as 'forbidden foods'.[27] The accompanying surge in dopamine, which is the result of the higher sugar and energy content as well as the result of caloric restriction, makes these foods much more rewarding to individuals in a state of energy depletion. This means that for anyone on a diet the temptations and the rewards offered by such 'forbidden foods' can prove almost impossible to resist.

Stress and Obesity

It's also important to understand the part that stress can play in overeating. Prolonged stress upsets the regulation of a vitally important pathway between brain and body known as the Hypothalamic Pituitary Adrenal (HPA) Axis – see Figure 9.1 – which affects both how much we eat and how that food is metabolised.

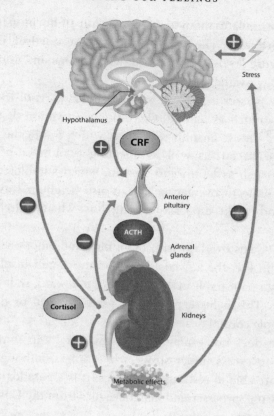

*Figure 9.1. The Hypothalamic Pituitary Adrenal (HPA) Axis. CRF
stands for Corticotropin-Releasing Factor and ACTH for
Adrenocorticotropic Hormone*

Stress causes the hypothalamus, described in Chapter 7, to
excrete a substance called Corticotropin-Releasing Factor (CRF).
This is transported through the bloodstream to the pea-sized
pituitary gland, which is attached to the hypothalamus by a
short stalk. There it stimulates the release of a particular
hormone, the adrenocorticotropic hormone (ACTH), and then
travels, again in the blood, to adrenal glands located above each

kidney. It stimulates them to secrete a group of hormones known as glucocorticoids. The most potent of these, cortisol, increases blood glucose levels and breaks down proteins and fats to help make energy available.

When a person is severely stressed, their cortisol levels may increase as much as tenfold. Continuous activation of the HPA axis contributes to insulin resistance, which forces the body to store sugar as fat, as well as affecting a number of hunger hormones such as leptin and ghrelin, which stimulate appetite and contribute to overeating. It can also heighten cravings for desserts and snacks, especially among those who are already over-weight.[28]

The HPA axis overlaps with the limbic and emotional regions in the brain (i.e. the amygdala, hippocampus and insula), all of which play a critical role in the experience of reward, and rewarding behaviour. This helps explain why, when stressed or depressed, many people comfort-eat HED foods.

What it does not explain is why this results in some people doing so to excess under stress, while others, although just as stressed, are able to restrain themselves. It is a paradox which, in 2013, Gudrun Sproesser and her colleagues from the University of Konstanz set out to solve.[29]

Munchers vs. Skippers

When stressed, four out of ten people (43%) seek relief by increasing their hedonic consumption of energy dense foods. These people are termed 'hyperphagics' or, in Sproesser's terminology, 'Munchers'. However, almost as many (36%) eat less when stressed – they are *hypo*phagics or 'Skippers'.

While accepting that Munchers did indeed eat more when experiencing negative emotions, Sproesser and her team wondered whether they might eat less than normal when in a regular, happy or otherwise more positive mood.

'Focusing on eating responses to negative situations might lead to the conclusion that stress hyperphagics are prone to over-consumption,' Sproesser comments, 'whereas a more comprehensive view, including responses to positive situations, might show adaptive eating behavior.'[30]

In the final mix of volunteers for Sproesser's experiment, one third (32%) were Munchers and the rest (68%) were Skippers. At the start of the experiment, each person watched a video featuring an agreeable individual of the same sex as the subject themselves, who described his or her interests, likes and dislikes. They were told that he or she might want to meet the participant – it all depended on how well they came across in a video they would make about themselves in return. After making this video and supposedly having had it viewed by the other student, they were randomly assigned to one of three groups, each of which was given different feedback as to how their video had been received.

Those in the 'social acceptance' group were informed the other student was looking forward to meeting them. Those allocated to the 'social rejection' group were told the other student did not want to meet up. The third, 'control' group was told that, unfortunately, no meeting could take place because the student had dropped out of the study.

Next all three groups were asked to taste three different ice creams. They had to report how much they liked them and whether or not they would buy them. Once the tasting session was completed, they were invited to eat as much ice cream as they wanted. As expected, the Munchers who had experienced 'social rejection' ate significantly more ice cream, while rejected Skippers ate significantly less in response to the same negative emotions.

Less expected and more interesting was the discovery that with 'social acceptance' and its accompanying positive emotions, this pattern of consumption was reversed. Now Munchers ate *less* ice cream and Skippers *more*.

The 'take home' message from this research is that emotional stimulation, whether positive and pleasurable (such as social excitement) or negative and painful (such as rejection), contributes to hedonic eating. It's understandable why this should be the case. The pleasure which the food provides is a form of escapism from a situation that is pushing us towards emotional extremes. Furthermore, eating in this way provides extra glucose to the brain to help it cope with a demanding situation. The unwelcome consequence is that the body is left to metabolise these excess calories. Moreover, if this strategy becomes habitual, the individual runs the risk of exhausting their insulin receptors, which will contribute to insulin resistance and increase the risk of developing diabetes.

The Hypoglycaemic Blues

So far in this chapter we have considered food consumption, especially hedonic eating, as a way of controlling emotions. But in some situations, food becomes not an answer – admittedly an unhelpful one – to emotional upsets, but their cause.

Imagine someone who starts their day at 7 a.m. by breakfasting on the following: coffee with full cream milk and two heaped teaspoonfuls of sugar; toasted white bread with strawberry jam, and a bowl of cornflakes. This satisfies their hunger at the time, but around 11 a.m. they start to feel not just peckish but also lacking in energy and a bit down. They have caught a dose of the hypoglycaemic blues. While the symptoms vary from one person to the next, these mid-morning blues typically include: anxiety, depression, lethargy, shakiness, irritability, dizziness, impaired concentration and daydreaming.

Looking back at the list of foods eaten at breakfast, it's pretty easy to work out the cause of these symptoms. All were medium to high on the Glycaemic Index (GI), which is a system for measuring how rapidly blood glucose levels are likely to rise after eating a particular food. Foods with a GI of less than 55 are rated as low; between 56–70 as medium and from 71 plus as high.

Since the speed at which blood glucose rises is influenced by the amount of carbohydrate in the food, it is also useful to consider a second factor, the Glycaemic Load (GL), which takes into account the amount of carbohydrate in the food and how much each gram raises blood glucose levels. In layman's terms, GI is how *fast* a food raises blood sugar, and GL is how *much* it raises it in total.

A GL of less than 10 is rated low, 11–19 as medium and greater than 20 as high.[31] Foods with a low GL will nearly always also have a low GI. Those with a medium-high GL can vary in their GI from very low to extremely high – watermelon, for example, has a high GI (up to 83) but, since a typical serving contains only a small amount of carbohydrate, has a low GL of around 4. Fructose (fruit sugar) has a low GI (11 per 25 g) and GL of only 1.

That said, foods with a low GI and GL can still cause a significant increase in blood glucose levels if they are consumed in large quantities. One lady interviewed during one of Margaret's investigations at the Kissileff Eating Lab at the University of Liverpool, for example, was putting on weight despite claiming that she ate only oranges. This seemed unlikely, until she admitted to eating more than thirty a day! While the amount of fructose in a single orange poses no problem, multiplied by 30 a day – 210 a week – it most certainly does!

With this in mind, let's take a look at the breakfast menu and note the GI and GL numbers of the food eaten.

Food	GI	GL
Sugar (sucrose) in coffee	58 medium	6 low
Toast (white wheat bread)	64 medium	8 low
Strawberry jam	61 medium	10 low
Cornflakes	100 high	23 high
Milk	40 low	4 low

In order to deal with all this sugar, the body produces increased amounts of insulin. The authors of *The New Sugar Busters! Cut Sugar to Trim Fat* liken insulin to a biological broom which: '. . . sweeps glucose, amino acids, and free fatty acids into cells where potential energy is stored as fat and glycogen to be used later.'[32] As we explained in Chapter 4 insulin is manufactured and secreted by beta cells in the pancreas. A healthy pancreas will secrete between 25 and 30 units (a unit here being 35 micrograms) of insulin each day and store around 200. Insulin works together with another hormone, glucagon, to maintain blood glucose at a healthy level. The former prevents it from rising too high and the latter ensures it does not fall too low. For these reasons insulin has been described as the 'feasting' and glucagon as the 'fasting' hormone. After insulin has done its work, the level of sugar in the blood declines rapidly and may even end up below its normal level. This is known as a 'sugar crash' or, as a doctor would describe it, 'hypoglycaemia' (see Figure 9.2).*

---

* This describes the condition in an adult whose blood sugar level is below 50 milligrams per decilitre (mg/dL), which is dangerously low and may lead to mental decline; dropping below 40 or 30 mg/dL can lead to seizures.

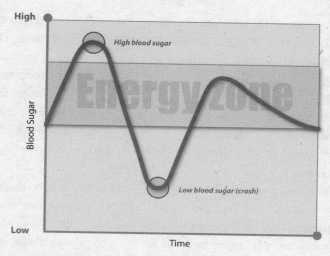

*Figure 9.2. Rise and fall in blood sugar levels
following a sugar-rich meal.*

If you have encountered these midday doldrums, you may have noticed that, as we described earlier, your first impulse is to eat HED snacks, in order to bring your blood glucose levels back up. Here are some of the treats which people typically consume for this purpose, together with their GI and GL figures:

Chocolate bars (per 50 g = GI 55:GL 14)
Bags of crisps (per 50 g = GI 58:GL 12)
Chocolate chip muffins (per 60 g = GI 58:GL 17)
Iced cup cakes (per 38 g = GI 85:GL 19).

Consuming these sorts of foods, high on the Glycaemic Index, results in another rapid rise in blood sugar levels, lifting the mood and restoring energy levels. Unfortunately this relief is short lived. The sugar spike is soon followed by a further crash when the body releases insulin to regulate itself once more. The result is that blood sugar and insulin levels fluctuate wildly and unhealthily.

For those with a weight problem, this roller-coaster ride of insulin spikes and sugar crashes is particularly bad news. High insulin levels increase the storage of sugar in liver and muscle as glycogen. Insulin also activates the enzyme lipoprotein lipase, which is responsible for removing triglycerides (a type of fat) from the bloodstream and storing in fat cells. It also inhibits lipase, another enzyme, whose job is to break down stored fats. The overall result is an increase in both fat storage and, inevitably, body mass, especially around the abdomen.

Unfortunately the bad news doesn't end there. When insulin is overproduced, both liver and muscle cells can become increasingly insensitive to it, and therefore resistant to normal levels. As a result, ever more must be manufactured in order to restore blood sugar levels to normal. This can create a downward spiral, ending with a failing pancreas and Type II diabetes.

And not only does insulin play a vital role in physiological health, but increasing evidence suggests it is also important in maintaining psychological health. In the elderly, for example, insulin sensitivity has been shown to have a positive correlation with verbal fluency, brain size, and the volume of grey matter in regions responsible for speech.[33] Insulin receptors are located in close proximity to the hippocampus, which is responsible for establishing long- and short-term memories, regulating mood, and also spatial orientation. For these reasons it has recently been proposed that insulin resistance (one of the hallmark signs of the development of diabetes) contributes to cognitive decline.[34]

With the biology and psychology of weight gain now described, we need to turn our attention to the world in which these systems function. A world which, as we have already explained, is very, very different from the one in which we evolved. It is a world beset with traps for the innocent, the unwary and the uninformed. A world in which the search for corporate profits is helping to make consumers fat.

PART FOUR

Our Obesogenic World

'The way ice cubes hit a glass, or even something as simple as a colour, which is totally unrelated to either food or drugs . . . these things can trigger an 'on' switch . . . Being in a perpetual state of "on" can be a problem in an environment characterized by excessive food, and food related paraphernalia.'

Kent Berridge[1]

CHAPTER 10

Caution: Food Cues at Work

'The (food) industry is geared to over-stimulating the senses of
the consumer so that they can eat more. It is activating parts of
the brain that are susceptible to being conditioned to find the
product desirable and wanting more of it.'
Gordon Shepherd, Professor of Neuroscience and Neurobiology,
Yale School of Medicine[1]

Whilst strolling along the street, your eye is suddenly caught by
the sight of a Starbucks on the other side of the road. Even though
you had not, until that moment, felt hungry, you are suddenly
seized by the desire for a coffee and chocolate cake. Images of a
steaming latte and a slice of mouthwatering gateau flash into your
mind. You can almost smell the aroma of freshly brewed coffee
and taste the rich, dark cake. A few seconds later, you are opening
the door of the shop and stepping inside.

It's not that you have a physical *need* for food, but rather that
the familiar Starbucks mermaid logo has created a psychological
want for it. It cued a memory of just how much you enjoyed coffee
and cake the last time you had them. The same might be true of
any number of other logos, the 'golden arches' of McDonald's
evoking the memory of a delicious burger or the blue and red
square of Domino's Pizza reminding you of the last time you
indulged in a feast of melted cheese.

'Those kinds of cues', comments Kent Berridge, Professor of
Psychology and Neuroscience at the University of Michigan, 'have
the power to evoke the desire to take that thing again.'[2]

Food cues are all around us. At home, we watch TV programmes and commercials discussing, describing and promoting a vast range of foods and beverages. We read newspapers and magazines filled with recipes and articles about growing, selling and consuming food. Away from home, we are bombarded with the sights, smells and tastes of foods. Billboards and advertising hoardings carry images of food and drink. Pavements are lined with cafés, bars, bistros, fast-food eateries, and fine dining restaurants. There are shops and vendors selling virtually every kind of snack from chocolates and sweets through to soft drinks, crisps, cakes, buns, pasties and hot dogs. For anyone living in a town or city, as soon as the desire to eat arises, it can be easily and instantly satisfied.

Central to this problem is what we might term the 'see food – eat food' dieting paradox. Those who most want and need to lose weight are the ones most sensitive to food cues. Their greater vulnerability, of which they are not usually consciously aware, arises for one main reason: they have learned to overeat.

Conditioned Consumption

During the early 1980s, Harvey Weingarten of McMaster University in Ontario, Canada, trained rats to eat on command. He did this by pairing the introduction of food into their feeding trays with either a light or a musical tone. The rodents quickly cottoned on to the fact that these cues indicated that lunch had been served. The rats had been 'conditioned' to associate a cue with eating. The interesting thing was that when Weingarten then allowed his rats to eat their fill at any time they wanted, the conditioning remained; even if they were completely satiated, if a light flashed or a tone sounded, the rats scuttled obediently to their feeding trays and gobbled up everything on offer.

'The results demonstrate that cues that have become signals for food can subsequently initiate a meal,' Harvey Weingarten comments. 'Once such an association has been learned, stimuli retain their ability to influence feeding for protracted periods and even under a state, satiation, that might have been expected to minimize the impact of such events.'[3]

Zombie Eating

Humans, too, can quickly and easily be conditioned to consume anything from food and drink to tobacco and drugs. All that it takes is for an association, at both a conscious and unconscious level, to be established between the stimulation of reward centres in the brain (as described in Chapter 7) and specific cues, or triggers, in the environment. These may be a place, a situation, an activity or a symbol. Some smokers, for example, light up almost without thinking about it when getting behind the wheel of their car, drinking coffee, opening a newspaper or in response to a tobacco advertisement. It's the same with those primed to respond to food cues. The sight of a logo on a shop front, or an advert flashing up on the television screen can create an impulse to snack on the associated food that is acted upon almost automatically.

If this kind of eating isn't really scrutinised by the conscious mind, the concern is that we are even less likely to be able to moderate it. We decided to investigate this. In a London cinema, we watched in fascination as an audience, deeply immersed in a Spiderman movie, ate popcorn like zombies. With large buckets resting on their laps, they munched their way through the snack with a methodical mindlessness. Without their eyes ever leaving the screen, their hands descended into the bucket and fistfuls of puffy white kernels were delivered to their mouths. Teeth and lips closed, jaws began chewing. Before that portion had even been swallowed, the hands began to reach down once more and the cycle was repeated.

So what would happen to popcorn consumption if that 'mindless' pattern was broken?

To find out we fitted each member of the audience with an oven glove worn on their dominant hand – the one they would usually use to pick up the popcorn. The results were pretty clear; when obliged to eat with their other hand, using an unfamiliar movement, the amount of popcorn people consumed fell by over a third.[4]

By slowing consumption and actually thinking about the food you are eating you can significantly reduce the number of calories you consume. Mindful – as opposed to mindless – consumption can prove an effective counter to the power of food cues.

Priming by Pizza and Ice Cream

Exposure to stimulus will often evoke subconscious thoughts of a second stimulus. The term 'priming' describes the phenomenon of how, for example, the word 'doughnut' is recognised faster if the word 'coffee' precedes it. Similarly, 'contextual cueing', a phrase coined by Marvin Chun and Yuhong Jiang of Yale University in 1998, refers to the way the brain seeks out previously identified patterns in its surroundings to identify new or unfamiliar stimuli. As with priming, the effect occurs in the 'implicit' or non-conscious memory. As a result, people remain unaware of the fact they are being influenced by, for example, repeated exposure to fast-food logos or familiar pack designs.

At the University of Florida in 2010, Carol Cornell and her colleagues conducted a study in which they explored the ability of two highly palatable foods, pizza and ice cream, to serve as food 'primes' likely to initiate eating and appetite.[5] They began by offering participants, young men aged from sixteen to twenty-eight, a free lunch and encouraging them to eat until they felt comfortably full.

After consuming around 1,000 calories each, the participants pronounced themselves satiated. They were then separated into two experimental groups and a control group. Those in one experimental group were asked to taste a mozzarella pizza and those in the second a bowl of vanilla and chocolate ice cream. Those in the control group were not given any food to taste. Members of both the experimental groups were instructed to take just one mouthful of the food in order to assess its appearance, flavour and aroma. Once this tasting had been completed, all participants were subsequently invited to eat as much as they wanted of the remaining pizza and ice cream.

The results were intriguing. Those who had tasted the pizza ate more pizza than ice cream, while those tasting the ice cream consumed more ice cream than pizza. On average they consumed an additional 300 kcal each, despite having reported only fifteen minutes earlier that they 'couldn't eat another thing'. This strongly suggests that everything further they ate was in response to a 'want' rather than a 'need'. Their preferences were the result of priming (by taking a single mouthful) and cueing (via the obvious availability of the foods).

'If you have a small taste of something, that can trigger overwhelming desire for that food. Our central problem is that appetite is a function of the circuits in the brain associated with reward; that circuit doesn't have a particularly strong "stop" signal,' explains Kent Berridge. 'So, in our world that circuit can be turned on and left on indefinitely. For some, that leads to issues with controlling our appetites.'[6]

In the previous study the participants, although left in the dark about the true purpose of the experiment, were otherwise aware of what was going on. However, priming need not always be so obvious. It can, in fact, occur without our ever being aware of the fact.

Subliminal Priming

During the summer of 1957, *Picnic*, a movie starring William Holden and Kim Novak, was doing roaring business for a cinema in Fort Lee, New Jersey. Billed as a love story between two people 'electrically attracted to each other' the film attracted more than 50,000 during its six-week run and won six Academy Awards. A couple of weeks later, a 42-year-old market researcher named James McDonald Vicary announced to journalists that those movie goers had been his unwitting 'lab rats' in a mind control experiment which secretly manipulated their desire to eat and drink. Vicary explained how he had installed a device of his own invention in the cinema which projected two advertising messages, one reading 'THIRSTY? DRINK COCA-COLA' and the other 'HUNGRY? EAT POPCORN' onto the screen as the film was showing.

Although these appeared continually throughout the movie, no one in the audience consciously saw them since they were flashed up for just three-thousandths of a second. Vicary reported that Coca-Cola sales had risen by 18% and popcorn by 58%. His claims created a storm of public outrage and universal condemnation by the media.[7] A few years later Vicary confessed it had all been a hoax, a PR stunt to drum up business for his near bankrupt 'subliminal projection' company.

For decades after this admission, advertisers vehemently insisted that they never had and never would employ such unethical tactics, while the majority of psychologists dismissed the possibility that subliminal messages could have any effect on behaviour.[8] But recent research has shown this confidence to be misplaced; in fact, subliminal cues can exert a significant influence over consumer choice and behaviour.

In one such investigation to determine whether unconscious priming could manifest itself in behaviour, scientists examined participants' reaction to emotional facial expressions and how much they drank after seeing happy versus angry faces.[9]

After asking whether or not they were thirsty, researchers instructed the participants to pour and then drink an unfamiliar beverage. They next completed a task that involved identifying whether a face on the computer screen was male or female. The expressions on these faces were either happy or angry.

Immediately after viewing the faces, participants were asked to rate their mood, and were then given the unknown beverage to drink, which was actually pleasantly flavoured with citrus. Both their self-reported mood, and how much of the beverage they consumed was recorded. While there was no detectable difference between participants' self-reported feelings, those who saw a greater number of happy faces consumed significantly more of the drink.

So it seems you can subconsciously prime people to drink or eat. If that is the case, is it also possible to influence exactly what they choose to eat or drink?

Thirsty? Drink Lipton's Iced Tea!

In 2006, Johan C. Karremans and his colleagues in the Department of Social Psychology at Radboud University, Nijmegen, set out to see whether it was possible to persuade thirsty volunteers to favour one beverage, Lipton's iced tea, over another, mineral water, by using subliminal priming.[10]

Participants watched a computer screen on which a string of capital letters, such as BBBBBBBBB appeared. Every so often, a lower case letter would occur, for example, BBBBBBBbBB. The job of the participants was to detect such differences. After completing this task they were given a salty local sweet, known as a 'dropje', to suck, and asked to try and identify a letter embossed on one side with their tongue. In fact this was just a cover story. The true purpose of the salty sweet was to increase their thirst. Next they were offered a choice of two thirst-quenching drinks, either a can of Lipton Ice Tea or a bottle of Spa Rood, a local brand of mineral water.

What those who had asked for the tea did not know was that they had been subliminally primed. While watching the strings of letters flow past them on the computer screen, either the words Lipton Ice or else nonsense control words containing the same number of letters, (e.g. Npeic Tol) had been flashed up for 23 milliseconds. Amazingly, 69% of those who had been subliminally exposed to the brand name chose Lipton Ice Tea, while only 25% of those presented with the control word did so.

'Our findings suggest that consumer choices may be influenced by subliminal primes,' comments Johan C. Karremans. 'Subliminal flashes of "Lipton Ice" on a television screen . . . may alter one's choice to order Lipton Ice.'

It's impossible to say whether subliminal advertising is currently being used to prime unknowing consumers to choose a particular brand of food or drink. While we have heard rumours that the technique has occasionally been tried, there is no firm evidence to support such claims. Nor is there any real need for this. Psychologists have demonstrated that behaviours can be just as effectively influenced by cues which, while visible to all, are rarely noticed by anyone.

Hidden in Plain Sight – the Power of Supraliminal Cues

While subliminal cues are presented too briefly for the conscious mind to be aware of them, *supraliminal* cues are easily seen, provided anyone attends to them. Supraliminal cues are simply items that you can register consciously, from the irritating jingle on the radio to the packaging colour of your favourite chocolate bar. If the item registers within consciousness, it is referred to as a 'supraliminal cue'. However, all the research evidence indicates that on most occasions people fail to register supraliminal cues, simply because their attention is elsewhere.

One of the first people to comment on this was the Hungarian neurologist and psychiatrist Resö Bálint. In 1907, he noted that: 'It is a well-known phenomenon that we do not notice anything happening in our surroundings while being absorbed in the inspection of something; focusing our attention on a certain object may happen to such an extent that we cannot perceive other objects placed in the peripheral parts of our visual field, although the light rays they emit arrive completely at the visual sphere of the cerebral cortex.'[11] In short, it's possible to look at something without consciously seeing it.

The ability of supraliminal cues to influence behaviour has been demonstrated by Esther Papies and Petra Hamstra from the University of Utrecht.[12] In their experiment they attached a modest-sized poster, displaying a recipe described as 'good for a slim figure', to the glass door of a butcher's shop. On the counter inside the shop they placed a tray of meaty snacks with a sign inviting people to taste as many as they wanted. One of the researchers stood nearby and surreptitiously counted how many snacks were eaten. The customers were then asked to complete a short questionnaire about their eating habits and say whether or not they were on a diet. After the poster had been up for a couple of days it was removed, but the snacks remained, and customers were still asked to complete the questionnaire.

Papies and Hamstra found that customers who were on a diet had seemingly responded to the 'slim figure' recipe – on the days it was up, dieting customers consumed fewer snacks than those not dieting. However, on days when the poster did not appear on the door, they ate significantly more than the non-dieters. The researchers reported that they believed their little poster, although seldom read or even glanced at, had: 'Overruled . . . the tendency to overeat on the presented snacks.' As David Kessler, former commissioner of the US Food and Drug Administration comments: 'Cues can gain power even if we're not consciously aware of them.'[13] Unfortunately, cues that help people lose weight are far less common in the modern world than those encouraging them to gain it.

We've mentioned just a few of the many studies demonstrating the power of subtle cues, of which we may not be consciously aware, to influence what and how much we eat. Even if this only involves us deciding to have a second helping of dessert, a bar of chocolate mid-morning, a doughnut with afternoon tea, or a snack when commuting, the calories quickly mount up. An extra 100 each day, the amount in a chocolate bar, will add 16 pounds of fat over a year.

How Food Cues Affect the Brain

Using functional Magnetic Resonance Imaging (as described in Chapter 7), scientists found that when shown images of high energy-dense foods, the brains of lean and obese individuals responded rather differently. The latter showed greater activation in regions associated with taste information processing (anterior insula and lateral orbitofrontal cortex), motivation (orbitofrontal cortex), emotion (amygdala) and memory (hippocampus).[14]

Food cues trigger activity within what is called the brain's 'appetitive network', comprising four major areas (described in Chapter 7): the amygdala, hippocampus, striatum, and orbitofrontal cortex. If the food cue is a visual one, a logo, billboard or TV commercial, for example, our eyes will send this information to the visual cortex at the back of the brain. From there, signals are sent to the hippocampus and the amygdala, regions responsible for memory storage and emotions. The hippocampus can simultaneously evoke a powerful memory and direct the individual towards a food reward, while the amygdala can endow that reward with powerful and extremely pleasurable emotion. Ingestion of hyper-palatable food will also excite activity within the striatum, the location of the nucleus accumbens, the brain's 'hedonic hotspot'. Once food high in sugar and fat is ingested, dopamine fires within two parts of the brain, in essence from the ventral tegmental area to the nucleus accumbens. It is this reaction that accounts for the hugely pleasant feeling we get when we eat hyper-palatable food. Finally, the orbitofrontal cortex (OFC) provides feedback, direction, and guidance for conscious action, leading us towards consumption or restraint from highly pleasant food. By the way, this entire process is happening almost below conscious awareness, simultaneous with the conscious pleasure of the 'Mmmm' when we bite into a delicious burger, or the, 'Those biscuits look nice,' when we pass them in the supermarket.

So, food cues alter activity in almost every part of the brain associated with motivated action. 'There is significant overlap with the regions in the brain that are associated with motivation and appetitive drive,' comments Kent Berridge. 'Once you initiate appetitive networks, in certain individuals you may also trigger motivational action. Since nice tasting food is almost everywhere, this can be a really difficult issue for people who struggle to turn the switch off.'[15]

In essence, seriously overweight individuals are far more sensitive to food cues than the majority. Not necessarily because they were born that way, but rather through a course of life-long learning. As with Harvey Weingarten's rats, mentioned earlier in the chapter, they have become conditioned to respond to even subtle food cues with an almost overwhelming desire to consume HED foods.

Wanting and Liking

Kent Berridge and Terry Robinson, from the Department of Psychology at the University of Michigan, have investigated the differences between wanting food and feeling hungry, what we termed homeostatic and hedonic motivation in Chapter 7.[16]

As we have explained, when we eat high energy-dense foods, dopamine and opioids flood the brain's reward centres, establishing or reinforcing powerful memories of the pleasurable experience. These can be stored as 'explicit' memories, of which we are consciously aware, but also as 'implicit', or subconscious, memories that can only be deduced from our behaviour. Whenever possible, advertisers and marketers seek to embed priming memories as deeply into the subconscious as they can. When they are able to do this, responses to their food cues become more reliable, since it is our implicit motivations – the ones of which we are not consciously aware – which are most intimately linked with our behaviour as consumers, as opposed to our explicit conscious ideas of what we should like.

'Liking' refers to explicit preference, the way we might say we like the colour blue. Liking tends to be a relatively innocent expression of preference. It is 'wanting' that causes so many issues with appetitive behaviour. While liking sets the stage for our actions and decisions, *wanting* determines our degree of motivation to buy something, or to eat it. Wanting is generated by the emotional centres of the brain, which include the nucleus accumbens and amygdala. For example, stimulation of the nucleus accumbens produces a voracious appetite in laboratory animals, especially for high sweet, high fat, foods.[17]

So it is clear that psychobiological factors affect our appetite. It's not simply that we think some foods 'taste nice', but rather that some foods, such as those high in sugar or fat, elicit a psychological, emotional feeling of pleasure. Both 'liking' and 'wanting' have an explicit, conscious aspect and implicit unconscious one; it is the implicit side which is the most problematic, tied up as it is with craving. As all dieters know, consciously deciding to go on a diet is the easy part. Resisting cravings is where things start to fall apart.

According to Berridge and Robinson: 'Wanting is a motivational, rather than an affective component of reward', which is to say that we may not be aware of how much we 'want' something until we feel that burn of craving it. It 'transforms mere sensory information about rewards and their cues (sight, sounds, and smells) into attractive, desired, riveting incentives.'[18]

It is known, for example, that repeat exposure to drugs can powerfully affect 'wanting' while 'liking' remains unchanged.[19] Within a very short space of time the 'wanting' can become so intense that it overwhelms any attempts at self-control. In the case of both drugs and food, this results in increasingly irrational and health-destructive behaviour.[20]

As scientists strive to make sense of this complex set of interactions, an important focus of obesity research is likely to be the hormones that act in concert with dopamine. This may, in time, lead to more effective ways of targeting overeating behaviour – perhaps even with the development of a weight-reducing pill. At the time of writing, new drugs that feature 'opioid agonists', such as naltrexone, have just been approved for use by the Food and Drug Administration. Naltrexone blocks opiate receptors, making highly palatable food less rewarding. While such pharmacological efforts have yet to have a significant impact in the battle against obesity, researchers remain hopeful of coming up with an effective drug.[21]

In summary, food cues contribute to triggering the motivational chain of events in the brain related to acquisition and consumption of food. Thus, they play a central role in explaining why it is that some people are more vulnerable than others to our 'obeso-genic' environment. What precisely these food cues are and how they trigger the desire to overeat will be explained in the next chapter.

CHAPTER 11

See Food – Eat Food: The Cues That Elicit Overeating

'We delude ourselves when we say that we are not influenced by
advertising, and we trivialise and ignore its growing significance
at our peril.'
Jean Kilbourne[1]

Food cues range from the obvious, logos, advertising posters and
television commercials, to the more subtle, including colours,
aromas and slogans. By using what is termed 'contextual cueing'
– that is delivering their consumption message at exactly the right
time and in precisely the right place – advertisers and marketers
are able to trigger a desire to eat at a non-conscious level. The
result is a sudden desire to eat which is overwhelming, impulsive
and typically mindless.

In this chapter we examine a dozen of the most frequently used
and potent cues, many of which you have probably already encountered several times today, perhaps without consciously being aware
of them.

1. The Persuasive Power of the Logo

The colours, images and symbols of company logos have become
a ubiquitous visual language in the modern world. These simple
graphics enable brands to be instantly identified and furthermore

require no translation, meaning companies can use them world-wide. Today, with thousands of brands competing for attention in an increasingly crowded commercial space, logos for food and beverages are among the most energetically and expensively promoted of the lot.

Take McDonald's, for example. With a current advertising and marketing budget in excess of $2 billion, McDonald's spends more money on getting itself noticed than four of America's other most popular fast-food chains combined.[2] A survey, conducted in the 1990s, found that while just over half (54%) those questioned worldwide were able to recognise the Christian cross, almost nine out of ten (88%) could identify the Golden Arches.[3] And Dr Emma Boyland, from the Institute of Psychology, Health and Society at the University of Liverpool, comments that children: 'Often learn the McDonald's symbol for M faster than the letter "M".'[4]

In an article entitled 'Love on a bun: How McDonald's won the burger wars', James Helmer from McMaster University concluded that the victory of McDonald's over its rivals in the fast-food wars of the 1980s was the result of the power of an advertising campaign which persuaded families that, beyond those golden arches they would discover: 'A potential source of love and human happiness . . . A place for *being* a family.'[5] This is an important point – that logos and other cues can affect behaviour, not just by triggering a desire for food in itself, but potentially by creating associations between that food and other qualities or attributes. For example, Chen-Bo Zhong and Sanford DeVoe from the University of Toronto found that participants in a study who were exposed to well-known fast-food logos became less patient, less able to slow down and less likely to save money. 'The consequences of fast food's ubiquity . . . are not adequately understood', they comment. 'The time-saving principle embodied by fast food can automatically induce haste and impatience.'[6] Nowhere is the cueing power of brand logos more apparent than in their influence over the very young.

According to author Juliet Schor, children are able to identify many logos before the age of two and will ask for products by their brand name at the age of three.[7] Also as early as three, children assert that food wrapped in paper printed with the McDonald's logo has a more pleasing taste; this holds true even when the food has not come from McDonald's. By eight, boys will start enjoying beer commercials (a firm favourite among this age group). At the same age, both sexes will have come to prefer fast food, high energy-dense snacks and sugar-laden soft drinks to home-cooked meals and less sweetened beverages.

While all children are vulnerable to food logos, research suggests those with an existing weight problem are especially at risk. When Amanda Bruce and her colleagues from the Hoglund Brain Imaging Center at the University of Kansas Medical Center used functional Magnetic Resonance Imaging (fMRI) to measure response to food logos in the brains of both healthy weight (HW) and obese youngsters, they found important differences between the two. Obese children showed the same level of activation in response to food logos *after* eating a satisfying meal as they did beforehand. This was not the case with healthy weight children, suggesting that the basic cause of obesity can be traced to dysfunctional hyper-responsiveness to food in early life.

'Obese youngsters failed to show significant post-meal reduction of activation in the prefrontal, limbic and reward processing (ventral striatum) regions,' Bruce explains. 'Whereas the Healthy Weight subjects showed a significant reduction in prefrontal and limbic activity after eating. These results indicate that the neural networks of the Healthy Weight subjects are more modulated after eating a meal.'[8] This means that the Healthy Weight children responded in a way more in line with a physiological, as opposed to psychological, motivation for eating. Their brains registered they had eaten, and therefore when viewing food cues showed decreased activation in areas related to eating motivation. The same was not the case for obese children, whose brains showed high levels of activation in response to food-related imagery, in both a hungry and sated state.

This research provides further confirmation that overweight children are hypersensitive to food cues at a very early age, and are therefore likely in the process of developing a diminished sensitivity to satiety signalling. Learning to eat beyond satiety is a pattern of food consumption that is carried into adulthood, making it a critical factor in the obesity problem.

2. As Seen on TV

In a landmark study of the power of TV food advertising to encourage overeating among small children, Jennifer Harris and her colleagues at Yale University divided 118 seven- to eleven-year-olds into two groups. One group watched a cartoon that included four 30-second food commercials, while the other group saw the same cartoons but without any commercials. The foods advertised were 'of poor nutritional quality using a fun and happiness message (a high-sugar cereal, waffle sticks with syrup, fruit roll-ups, and potato chips) and were chosen to represent the types of food commercials that are most commonly shown on children's TV.'[9] Each child was handed a bowl of cheese crackers and told they could eat as many as they liked while watching. After the children had left, researchers weighed the remaining crackers and found that those who had seen the food commercials consumed almost half as many crackers again as those who saw the cartoons without any commercials.

'Food advertising that promoted snacking, fun, happiness, and excitement, i.e. the majority of children's food advertisements, directly contributed to increased food intake,' Harris summarised. 'These effects occurred regardless of participants' initial hunger, and amount consumed after viewing snack advertising was completely dissociated with participants' reported hunger.'[10]

Young children are especially vulnerable to food commercials because they are less able to appreciate the persuasive intent of advertisements, naively regarding them accurate and trustworthy sources of information about the world.[11] Bright primary colours, which are particularly attractive to children, are significantly more likely to feature in food commercials aimed at this demographic. The advertisements are also scripted, photographed and edited to match the cognitive abilities and emotional needs of pre-teen viewers.[12] The good news for parents and health professionals is that, under pressure from regulators, advertising high energy-dense foods to children is far more restricted than in the recent past.

The bad news is that advertising for these products is migrating away from TV to mobile devices, such as phones and iPads. Here the marketing messages can be made far more powerful and much less easy for adults to monitor and control. We will examine this phenomenon in greater depth shortly.

3. The Power of Cartoon Cues

One of the ways advertisers try to match their content to the mental and emotional maturity of a young audience is by using cartoon characters to promote products. Apart from being associated with enjoyable TV shows, cartoon characters can be drawn in such a way to connect very directly with a small child's sense of wonder and desire for companionship. Aside from TV commercials, the place children are most likely to encounter cartoons is on boxes of breakfast cereals. Cereal companies spend more on marketing their products to children than to any other group of consumers.[13]

In the US, the cereal industry spends a total of $3 billion annually on packaging targeted at children. Since the fronts of cereal boxes are especially important to attract young consumers they almost invariably feature brightly coloured cartoon characters known in the trade as 'spokes-characters'.[14]

These graphics both aid brand recognition and communicate fun, enjoyment and trust. Unfortunately, while having a colourful cereals package on the table may make breakfasts more fun, they are not the most nutritionally sound way to start the day. Researchers have found that eight out of ten of the least nutritious cereal brands feature cartoon characters, compared to just two of those rated 'most nutritious'.[15] Cereals advertised to children have, on average, 57% more sugar and 50% more salt than adult brands, but 52% less fibre.[16] By the age of two, most children can recognise a variety of cartoon characters, and will have formed a strong emotional bond with their favourites.[17] They identify closely with the cartoon and aspire to emulate them.[18] When given identical foods to taste, where one features a familiar cartoon character on the packaging and one does not, children believe that the food with the cartoon tastes better.[19]

Adding to the power of some spokes-characters as food cues, is their longevity. Tony the Tiger, who appears on packets of Kellogg's Frosted Flakes, was created by graphics artist Eugene Kolkey in 1951. The sweetened corn and oats breakfast cereal fronted by Cap'n Crunch dates back to 1963. As a result, parents and grandparents may experience a sense of nostalgia when seeing them on the supermarket shelves. This helps generate trust in the brand and makes it more likely they will encourage their children and grandchildren to eat this cereal, since they did so themselves.[20]

And it seems that some spokes-characters are eye-catching in a far more literal way than simply through being bright and colourful. When Aviva Musicus, from the Yale Rudd Center for Food Policy and Obesity, and Brian Wansink, from the Food and

Brand Lab at Cornell, examined more than 80 different cartoon characters on cereal brands aimed at children, they found something quite remarkable. The gaze of these characters seemed to make direct eye contact with a small child looking up at the box on a supermarket shelf. When someone meets our gaze we subconsciously evaluate them as friendlier and more open, honest and trustworthy. If, on the other hand, they avoid our gaze we tend, again often subconsciously, to view them as unfriendly, dishonest and untrustworthy. Remarkably, the same seems to hold for cartoon characters on cereal packets.[21]

In their report, entitled 'Eyes in the Aisles: Why is Cap'n Crunch Looking Down at My Child?', Musicus and Wansink described how 57 of the cartoon characters were drawn with a downward gaze of just under 10 degrees. This meant that when displayed on shelves slightly higher than a child, they made direct eye contact. By contrast, characters on cereals marketed at adults gazed upward at around 0.43 degrees. These boxes were, typically, placed on the top two shelves at a height of around 54 inches, once again perfect for making eye contact, but this time with the average grown-up shopper. Musicus and Wansink also examined the extent to which eye contact with cereal-box characters influenced feelings of trust and connection with a brand. University students were randomly presented with one of two versions of a cereal box featuring a cartoon rabbit. On one, the animal looked directly back at the viewer, while in the second its eyes were shown looking down. The results showed that the eye contact rabbit increased trust in the brand by 16% and the feeling of connection by 28%.[22]

The study attracted a lot of attention, generating a heated debate on social media and a firm rebuttal from cereal manufacturers and even a few criticisms from academia. Tom Forsythe, vice president of Global Communications for Minneapolis-based General Mills, wrote on his blog: 'When should companies respond to poor research – and pseudo-science? . . . Sometimes it's best to just let stuff go . . . this one is getting a response — because it's absurd.'[23]

What no one disputes is that cartoon spokes-characters are a hugely effective weapon in the battles between rival breakfast cereals. They are one of the main reasons children develop a taste for a particular cereal in the first place, and by establishing a connection with children at a young age, cereal makers, whether deliberately or not, are creating a connection with these cartoon characters that will shape consumers' tastes for the rest of their lives.

In recent years there have been both positive and less positive changes in cereals marketed to children. On the plus side, the Yale Rudd Center for Food Policy & Obesity reports that – between 2008 and 2011 – the overall nutritional quality of 13 out of 16 brands targeted at children increased by an average of 10%.[24] Ten out of twenty-two brands examined reduced sodium and seven reduced sugar, while five increased fibre. Some child-targeted websites and games were discontinued and one major company, General Mills, reduced banner advertising on children's websites by 43%.

However, 'Total media spending to promote child-targeted cereals increased by 34%, from $197 million in 2008 to $264 million in 2011,' report Jennifer Harris and her Yale Rudd Center colleagues.[25]

Harris also states that in 2011, the number of child visitors increased for eight out of ten child-targeted cereal websites with an average of 162,000 children a month visiting Kellogg's frootloops.com and 116,000 children a month visiting applejacks. com. Companies also increased the amount of banner advertising on children's websites, in one case by 185% (for Honey Nut Cheerios).

Less encouraging was the finding that advertising to children for many of the least nutritious cereals still rose substantially overall. While these are American findings, a similar pattern of results could reasonably be predicted in other developed nations.

4. Giving Obesity a Sporting Chance

In the world of marketing HED foods, sporting icons and exhortations to exercise reign supreme. On the surface this might seem counterintuitive, but of course there's a very good reason for it; by priming customers to associate a particular soft drink, chocolate bar or energy drink with sport, manufacturers are able to cue consumption through subconscious associations with fitness and health. This is why so many food companies sponsor major sporting events, including the Olympics. Some even host conferences on nutrition. They are effectively giving obesity a sporting chance.

Marie Bragg and her colleagues from the Rudd Center for Food Policy and Obesity at Yale University examined more than a hundred products which featured some kind of reference to sports on supermarket shelves (fifty-three foods and forty-nine beverages). They found that the vast majority (73%) depicted people taking exercise on their packaging.

Four out of ten products (42%) carried an endorsement by at least one professional athlete, elite sports organisation or well-known sports team. A third offered the chance to win some kind of prize. Three-quarters depicted at least one type of sports equipment, while one in ten (9%) were manufactured to resemble sports equipment; for example the special edition football-shaped Oreos. Finally, taking us back to the issue of priming consumers at a young age, a third of the products were targeted specifically at children, featuring either a cartoon character on their packaging or a word synonymous with 'child'.[26]

None of these 'sporty' foods were especially nutritious, while many were downright unhealthy. More than two thirds of the beverages, for example, were 100% sweetened with sugar. Rated on a scale from between 1 (unhealthiest) and 100 (healthiest), the median score of these randomly selected products was just 36.

5. Celebrity Endorsements

Celebrities are paid handsomely to endorse a wide range of food products, companies and retailers. Having a brand associated with a well-known sportsperson, television personality or film star is an extremely effective way of creating recognition, enhancing credibility and adding value.

In a second study by Marie Bragg and her colleagues in 2014, it was found that around four out of every five food endorsements by influential athletes involved products high in calories but low in nutrients. After analysing TV, radio, newspaper and magazine endorsements by the top 100 professional athletes, they reported that 79% of the heavily promoted food products offered few nutrients and had high calorie counts. In just over 93% of the endorsed drinks, added sugar accounted for all the calories provided.[27]

'The promotion of energy dense, nutrient-poor products by some of the world's most physically fit and well-known athletes is an ironic combination that sends mixed messages about diet and health,' Bragg comments. 'It is possible that food companies associate with athletes simply because they are celebrities, but research shows that athlete endorsements are associated with higher healthfulness ratings on the products they endorse.'[28]

A case in point from the UK is the twenty-year endorsement of Walkers Crisps by former England international footballer Gary Lineker, now a popular sports commentator. Emma Boyland and her colleagues from the University of Liverpool conducted research into the impact such endorsements have on children.

Boyland asked 181 children, aged between 8 and 11 years old, either to watch a 20-minute Simpson's cartoon, into which was embedded one of three different advertisements (one for Walkers Ready Salted Crisps, endorsed on screen by Gary Lineker; one for a brand of salted peanuts, Nobby's Nuts; and one for a non-food-related toy), or to watch a programme featuring highlights from *Match of the Day* with Lineker as the main presenter. The children were also provided with two bowls of crisps, one labelled 'Walkers' and the other 'Supermarket' – although both actually contained Walkers Crisps.

At the end of the session the quantity of crisps remaining in each child's bowl was recorded. It was found that the children who saw the commercial for Walkers Crisps endorsed by Gary Lineker, or the football highlights programme which he presented, both consumed considerably more of the Walkers Crisps than the children who watched either the Nobby's Nuts or toy advertisements. So the experiment strongly suggests that celebrity endorsements do encourage increased consumption by children.[29]

'The study demonstrated for the first time, that the influence of the celebrity extended even further than expected and prompted the children to eat the endorsed product even when they saw the celebrity outside of any actual promotion for the brand,' commented Boyland. 'If celebrity endorsement of HFSS (High Fat Salt and Sugar products) continues and their appearance in other contexts prompts unhealthy food intake then this would mean that the more prominent the celebrity the more detrimental the effects on children's diets.'[30]

'There's no doubt,' says Yoni Freedhoff, an obesity expert and assistant professor of family medicine at the University of Ottawa, 'that the presence of their [children's] heroes, heroes who themselves are the embodiment of physical fitness and health, serve to very clearly "health-wash" the products.' Freedhoff observes that the least nutritious foods tend to be the food industry's biggest profit drivers and that celebrities and athletes help to fulfil commercial pressures to maximise profit. [31]

6: Health Claims

Paradoxically, the rise in obesity has recently been matched by an increasing emphasis on the importance of healthy eating, with products being heavily marketed as low in fat, sugar and salt. Such claims both increase sales and lead consumers to overestimate just how healthy the product actually is.[32] For example, a reduction in fat may be accompanied by an increase in sugar to compensate for the reduction in taste. If someone is under the impression that a product is 'healthy', they may well feel free to consume more of it than they would otherwise.

7: Colour Me Delicious

By associating their brand with a specific colour, food companies can spark brand recognition faster than is possible with even the slickest slogan or brand name. The brain takes far less time to recognise a colour than interpret a text. For example, Bailey Dougherty, account director at Boom! Marketing, a leading Canadian Experiential Marketing Agency, believes that the net worth of the colour purple to Cadbury's is 'almost priceless,' since 'it is associated with a series of feel good emotions. Cadbury Purple is regal yet accessible – and you may have noticed, that seeing the colour does make you think of chocolate.'[33] In the world of modern advertising even something as apparently general as a colour can be imbued with specific associations which turn it into a food cue.

8: The Smell of Success

The next time you visit a supermarket or department store, have a good sniff. It's increasingly likely that you will catch a whiff of an aromatic food cue designed to whet your appetite and encourage you to spend more money. Some stores use just one scent, such as freshly ground coffee or newly baked bread, while others use a sophisticated variety; in a Net Cost Market in New York, for example, the aroma of chocolate hangs in the air around the sweet snacks displays, the scent of grapefruit wafts through the produce department and you can smell rosemary focaccia in the bakery section. Since introducing these aroma cues, the store reports that sales have risen by almost 10%.[34]

Nor are aromatic food cues only found inside stores. You may even encounter them in bus shelters and, in some parts of the world, on commuter buses. In 2012, McCain foods used a combination of heat and aroma cues to boost sales of a new range of microwaveable baked potatoes. The campaign involved fitting UK bus shelters with heaters disguised as jacket potatoes; when a waiting passenger pressed the button to get some warmth, the shelter was filled with both hot air and the aroma of freshly baked potatoes. Because the promotion was launched during a miserably cold February, the heaters proved extremely popular with the shivering public. The warmth and the accompanying smell no doubt evoked memories of the comforts of home and the enjoyment of eating a freshly baked potato. These cues were apparently irresistible to many people, since McCain's sales increased significantly.[35]

A somewhat similar approach was used in South Korea, by Dunkin' Donuts. They introduced the smell of freshly brewed coffee into buses whenever their advertisement was played on the radio. This reportedly increased the number of customers visiting their stores by 16% and sales by 29%.[36]

9: Slogan Power

Many food brands use words and phrases either to cue a desire to eat at that precise moment or else to foster a notion that by consuming their particular product one can become healthier, fitter, more energetic, happier and so on. They focus on such basic human motivations as the need to feel loved, wanted and socially accepted; by marketing a food or beverage as being able to help consumers meet one of these needs they can significantly increase that product's attraction.

Copywriters strive to produce memorable phrases that are effortlessly brought to mind when triggered by a particular location, occasion, emotion or even the time of day. McDonald's famous slogan, 'I'm lovin' it' (from the Justin Timberlake song of the same name written by Pharrell Williams), was dreamed up by Paul Tilley for the advertising agency DDB in 2003. It proved so universally successful that it has been translated into more than twenty languages.

Notice that it's 'I'm lovin' it', rather than 'You're lovin' it'. Telling people to behave in a certain way risks generating psychological resistance that can actually cause them to do the exact opposite. But the easily remembered phrase 'I'm lovin' it' is personal without being instructional. It can echo in your head as a form of autosuggestion, subtly persuading you that, yes, you really do 'love' McDonald's.

'Emotional states are trance states, and memories are state dependent. That means that if a company can link their brand with both an emotion and with a real world trigger, not only will that trigger remind a consumer of the brand, but it will also make them more likely to feel that emotion. And potentially even more powerfully, when the consumer feels that emotion in other contexts, they will also be more likely to think about the brand,' explains hypnotherapist Dan Jones.[37]

10: Digital Marketing

Before leaving the topic of food cues, we need to consider how digital technologies are transforming food marketing, particularly to children. While TV is still significant when marketing to younger children, for teens and increasingly pre-teens, television has started to take a back seat to digital media. A US survey published in 2013 found that almost four out of ten children (38%) under the age of two were familiar with mobile phones and tablets.[38] By the time they start primary school, 70% of that cohort have mastered them completely. Before their eighth birthday, 72% of children will have used a mobile device for playing games or watching videos.

Talking about the relationship many children now have with their mobile devices, Steve Smith, editor of *Mobile Marketing Daily* says, 'They say they will never leave home without it. They even report feelings of withdrawal from leaving their phones at home.'[39] Mobile technology is at the centre of the lives of many children and teenagers.

Never before has so much personal information been available to food and beverage companies, and digital marketing offers one of the broadest and most far-reaching platforms in history. The extensive amount of personal information we give on social media is incredibly valuable to companies, so much so that they are willing to buy it. They are then able to craft bespoke cues that cater specifically to an individual's likes and dislikes.

For example, the more you engage with certain restaurant chains (say, by checking in via Facebook), the higher the probability that you will receive pop-up promotions from this organisation on your news feed.

And it doesn't end there. Global Positioning Systems (GPS) mean that the physical location of a device can be taken advantage of for marketing purposes. Two students in the laboratory of Stanford University computer specialist Brian Fogg came up with a clever way of exploiting this technology. They designed a cuddly toy bear that a fast-food chain could either give away to children or sell at a low cost. Whenever the bear came close to a branch of that restaurant chain, it would start to sing a jingle about how delicious French fries are and how much it enjoyed eating them. Fortunately for the sanity of parents, the bear never got beyond the prototype stage. Had it ever been put into production, its potential for causing small children to demand a burger could have been considerable, though one imagines it would also have generated significant public backlash.[40]

However, given the commercial potential of digital marketing it comes as little surprise that companies like Coca-Cola, PepsiCo, McDonald's, Kellogg's, and Burger King are leading the way in developing other location-marketing technologies. Increasingly, companies are designing their websites to run on mobile phones and are encouraging users to download apps which tell marketers their location – this information can then be used to target them with specific messages and offers.

For example, a person walking down the street could be sent a text offering a 25% discount, a free serving of fries or a supersized coke – an incentive to call in at their favourite fast-food outlet. The text could even point them in the direction of the closest branch. Such suggestions are especially likely to prove of interest if deliberately timed to coincide with the mid-morning or mid-afternoon dip in energy many people experience.

Known as SOLOMO (Social – Local – Mobile), this type of digital marketing is the latest way to attract customers and cue them to eat more food. 'So-Lo-Mo isn't a fad that's going away,' says writer Lindsay Scarpello. 'It's here, it's real, and it's important for brands and retailers, as well as marketers, to get hip to this trend.'[41]

11: Playing Games with Obesity

Where teenagers and young adults are concerned, one of the most powerful and effective ways of spreading the word about a food brand is to catch their attention while they are playing online games.

Computer games first arrived on the scene with the invention of Pong during the early 1970s. Forty years on, and the use of high-definition graphics, motion tracking and other technologies is blurring the distinction between games and reality. Players are invited to immerse themselves in a fantasy that stimulates all their senses, which today can even include smell and touch. Such total immersion in an emotionally arousing virtual world undermines self-control by depleting the mental resources needed to inhibit impulsive behaviours. As we saw in Chapter 8, impulsive eating is one of the key causes of overconsumption, so anything which makes it harder to resist such impulses is a potential health problem.

However, the issue is not only the potentially negative consequences of playing games in and of themselves. Games also enable companies to integrate advertising and product placements into the action, and to do this so naturally that, for the average player, these messages become an integral part of the game world itself.

In 2013 Kellogg's released the first food-company, child-targeted 'advergame' – an app called 'Race to the Bowl Rally'. It could be downloaded for free and played on a computer, mobile phone or tablet. The game involved racing through a frozen landscape and collecting pieces of Kellogg's Apple Jacks cereal along the way. With a nutrition score of 44, in 2012 this cereal was described by the Yale Rudd Center for Food Policy and Obesity as among the worst on the market in terms of sugar content and absence of fibre.[42]

Race to the Bowl Rally is just one of a number of advergames used by cereal companies to promote their products to children. In 2011, Post and General Mills launched PebblesPlay.com, HoneyDefender.com (promoting Honey Nut Cheerios) and CrazySquares.com (promoting Cinnamon Toast Crunch).[43]

Children who play such games will quickly come to associate the fun of playing them with the brand and its products. And there are even more sophisticated forms of branded online gaming out there. Four years before Race to the Bowl Rally, Frito Lay launched a scary Halloween-themed game targeted at tweens and teens. Their marketing objective was to 'bring back from the dead' Doritos Black Pepper Jack and Smoking Cheddar, two flavours that had fallen out of favour with younger consumers.

The game they devised was called Hotel 626. In it, players found themselves trapped in a haunted hotel, from which they had to escape by completing a series of scary challenges involving webcams and mobile phones. A live Twitter feed enabled players to share their experiences and they were encouraged to 'send a scare' to friends on Facebook. To add to the 'scare quotient' the game could only be played after dark. This combination of real and virtual realities produced intense emotions, while simultaneously promoting the consumption of a highly palatable snack. The game was effectively priming its youthful players to develop a powerful emotional connection with Doritos.

Costing less than $1 million to launch, Hotel 626 proved an overwhelming marketing success. Within a few weeks, over four million young people from more than 136 countries had checked in and played for an average 13 minutes. Over 2 million bags of the relaunched flavours were sold within three weeks.

Two years later, an even more sophisticated version of the game was released. In Asylum 626, players were chased through the corridors of a mental asylum, by nurses wielding chainsaws. This game differed from the first in that players were able to physically interact with the game world.

'We employed head tracking in one scene, so the player literally must move to avoid an attack,' explained Hunter Hindman, the campaign's creative director. 'We used the webcam in new and innovative ways to actually place the player into the gameplay itself. We asked people to give us more access and information this year, telling them upfront that the more they gave us, the scarier the experience. We used social networking in ways that hadn't been done before. Specifically, we bring their friends into the experience and the gameplay itself. All of these changes began to add up to . . . a more immersive, more frightening experience.'[44]

In order to finally make their escape from the asylum, players had to present their webcam with a marker printed on bags of the two flavours being promoted, meaning that completing the game was contingent upon making a purchase.

Both of these games appealed to adolescents. This is a significant point, since the brain's prefrontal cortex, which plays a critical role in decision-making, does not reach full maturation until early adulthood. As a result, tweens and adolescents are especially vulnerable to fear-related stimuli and far more receptive to the rewarding properties of HED food.

Researchers at the University of California Irvine report that a game of this type 'purposefully evokes high emotional arousal and urges adolescents to make consumption decisions under high arousal,' exacerbating their tendency for poor decision-making when emotionally aroused, and also their tendency to self-medicate using foods high in sugar and fat.[45] These findings confirmed an earlier study by Microsoft, which reported that such campaigns: 'Evoke stronger emotional connections with consumers and more positive emotional association with the brands.'[46]

12: Music Cues

The power of music to trigger emotion is well known – whether that emotion is joy, depression, or spine-tingling excitement. This, too, is something that can be taken advantage of to market food products. Just a few bars of music – known as a 'sonic signature' – can be enough to trigger a memory and bring a brand to the forefront of a consumer's mind, in a process known as 'involuntary music imagery' or INMI. INMI is defined as 'a conscious experience of reliving a musical memory without deliberately attempting to do so,' explains Dr Lassi Liikkanen, a researcher at Aalto University, Finland.[47]

Although the reasons INMI occurs remain unknown at the time of writing, it's hard to refute that the more often one hears a simple piece of music, the more likely it is to stick in one's head. It can be almost impossible to banish from the forefront of one's mind, at least for a while. Which may be irritating for the consumer, but it is literally music to the ears of marketers, advertisers and retailers.

When listening to music the brain's auditory, motor and limbic (emotional) regions are simultaneously engaged. This helps to ensure that an associated visual cue, such as a logo, is remembered for longer and in much greater detail because it is being encoded at a more basic and primary part of the brain (the limbic system). Music cues also work even when consumers are not consciously aware of them. For example, Charles Gulas, an associate professor of marketing at Wright State University, and Charles Schewe, professor of marketing at the University of Michigan, found that baby-boomers were more likely to spend money in shops playing classic rock, despite the fact that 75% of customers were unable to recall what type of music they had heard in the store when asked afterwards.[48]

Similarly, when classical music, as opposed to tunes from the Top 40, was played in a supermarket's wine department, customers purchased more expensive wines.[49]

Music with a rapid tempo encourages diners in fast-food restaurants to eat quickly, while slow, classical music in a high-class restaurant slows down the pace of eating and encourages diners to linger. Paradoxically, we may eat less when encouraged to slow down and enjoy the ambience. That means that pop music played in a fast-food restaurant may well result in the rushed consumption of greater quantities of food.

Before leaving the topic, it is worth mentioning two subtle, attention-grabbing forms of audio branding known as 'Earcons' and Icons. Earcons were originally designed in the late 1980s as audio messages 'used in the user-computer interface to provide information and feedback to the user.'[50] For example, the sound your computer makes when you turn it on, or the generic ring applied to all iPhones when first bought. Today the term is applied to any brief series of notes, such as the 'Bah-da-ba-ba-baahhhhh' used by McDonald's in their 'I'm lovin' it' campaign, mentioned earlier.

Auditory icons, on the other hand, use not music but everyday sounds. These aural associations have to be learned, but once established they can serve as powerful food cues. You might, for instance, hear the sound of a steak sizzling when you click on a restaurant's icon or the hiss of an espresso machine as you pass a coffee shop. Both could trigger a sudden desire – to visit that restaurant or to have a coffee.[51]

Manipulated Man?

In his 1924 book *Behaviourism*, psychologist John Broadus Watson boasted that given a dozen healthy infants and his own world in which to raise them, he would: 'Guarantee to take any one at random and train him to become any type of specialist I might select regardless of his talents, penchants, tendencies, abilities, vocation, and race of his ancestors.'[52]

This pretty much sums up the attitude of those whose business is to target today's food consumers, especially the younger and more easily impressionable ones. However, their interest is not in guiding the careers of the upcoming generation, but rather in controlling their appetites; implanting food cues within their subconscious that will influence their eating habits throughout life.

CHAPTER 12

Whatever Have They Done to Our Food?

'Our food system is an hourglass. In one chamber are tens of thousands of farmers and ranchers, but their sands are steadily receding. In the other are hundreds of millions of eaters, whose sands continue to swell. In the narrow middle between growers and eaters sit a handful of giant corporations, what economists call an oligopoly.'

Aman Singh, *Forbes* magazine[1]

Walk into any major supermarket, almost anywhere in the world, and you'll be confronted by a cornucopia of choice. Shelves are filled to capacity with cereals, drinks, snacks and desserts; displays are stacked with cartons, cans, bottles, bags and packets of food; boxes overflow with fresh fruit and vegetables flown in from all over the world; counters are piled high with poultry, fish and meat.

When Barry Schwartz, Professor of Social Theory and Social Action at Swarthmore College, visited his moderately sized local supermarket in the US, he counted: '85 different varieties and brands of crackers . . . 285 varieties of cookies . . . 13 "sports drinks", 65 "box drinks" for kids, 85 other flavours and brands of juices . . . 29 different chicken soups . . . 16 varieties of mash potatoes . . . 120 different pasta sauces . . . 275 varieties of cereal . . . and 175 types of tea bags.'[2]

While this superabundance of choice might seem sufficient to satisfy even the choosiest consumer, the range of products available from local stores pales into insignificance compared with the 50,000 or more products on offer at a major supermarket. You can find three times as many products as were available thirty years ago, yet still new ones arrive in their hundreds each and every month. According to Donald Stull, from the University of Kansas, and Michael Broadway, from Northern Michigan University: 'In 2010 alone, more than 15,000 new foods and beverages came to market in the US. But such is the competition for our food dollar that many of these new products will fail. So much to choose from, no wonder we usually come home with stuff that wasn't on our shopping list.'[3] Yet this unprecedented range of options, this seeming ability to satisfy each and every consumer's individual desires, masks the fact that the number of sources from which our food now comes has declined sharply over recent years. This has led some more discerning customers to ask in bewilderment: 'Whatever have they done to our food?'

In this chapter we aim to address that question and to describe both the benefits and the less desirable consequences of the way food is now grown, processed, marketed and sold.

The Changing Nature of our Food – From Farming to Fordism

In their book, *Food in Society: Economy, Culture, Geography*, Peter Atkins from the University of Durham and Ian Bowler from the University of Leicester identified three successive 'food regimes' between the mid-1800s and today.[4] Each one is characterised by very different forms of production, marketing and retailing.

Regime 1 – the Farmers' Market

Up to the early 1920s, agriculture in the UK, Europe and much of the US was based on small (typically family run) farms, with much of the produce sold locally, often within a few miles of the farm gates. The invention of the mechanical reaper in the 1840s meant that farmers could harvest five to six acres daily rather than the two acres previously possible, enabling them to move from a modest level of production to commercial agriculture.[5]

The construction of refrigerated ships during the 1880s meant that distance was no longer an obstacle to transporting perishable foods, such as butter, meat and fruits. As a result, during the late nineteenth and early twentieth centuries, increasing amounts of food were imported from Africa, South America and Australia. Here, the land was owned, and production mostly controlled, by white farmers who employed native labour on low wages and frequently under appalling conditions.

Farm yields peaked during the first half of the twentieth century, only to be obliterated in the USA by the Dust Bowl and, on both sides of the Atlantic, by the Great Depression. These brought the inefficiencies that plagued this method of food production and distribution into sharp focus and led to the development of the second agricultural regime.

Regime 2 – the Rise of the Agri-Business

After World War II, American concern for feeding the poor led to significant increases in agricultural production, dubbed the 'Green Revolution'. Dr Norman Borlaug, known as the Father of the Green Revolution, dedicated his early post World War II career to improving wheat production. His goal was to develop a strain of wheat that was thick stemmed, resistant to disease and gave a higher yield than the varieties then available.[6]

The new semi-dwarf wheat he developed was such a success that by 1963, 95% of Mexico's crops were varieties of this, and the average yield was six times greater than twenty years previously. Borlaug won the Nobel Peace Prize, the Presidential Medal of Freedom, and the American National Medal of Science for his work, which has had a profound effect on our ability to provide sufficient food to the developed and developing world.[7]

Intended to aid the poor, and those hardest hit by droughts and crop failures, the Green Revolution was in many ways a major achievement. It takes significantly less land, and less energy to grow the same amount of food; for example, in the eighteenth century it would have required almost five acres of land to feed one person for one year. By the 1980s it took just half an acre.[8] Millions were lifted out of poverty and famines were averted, with developing nations enjoying a boost in food security as the Western world transferred the knowledge from its scientific and industrial revolutions into agriculture.[9] It is estimated that Borlaug's practices saved over a billion lives.[10]

However, the results of the Green Revolution were not uniformly positive. The fusion between science and agriculture paved the way towards new practices in irrigation and the increasing use of chemical fertilisers and pesticides. Many scientists believe that these developments have brought with them a host of new problems, perhaps eroding many of the so called benefits. Indian physicist Vandana Shiva, for example, dismissed the Green Revolution as a hoax by Big Agriculture, simply a vehicle for large agribusiness to exploit soil and capitalise on poor farmers. This has led her to suggest that these techniques have left Indian crops more vulnerable to drought and reliant on 'poisonous' agrochemicals.[11] When Borlaug tried to employ similar strategies in Africa in the 1980s, the World Bank and Ford Federation had to reduce funding as a result of pressure from environmental lobbyists.[12] Environmental lobbyists often cite the fact that business interests will complicate agricultural practice, and will destroy biodiversity and soil quality in the process.

Moreover, while the Green Revolution was originally aimed to address the food production needs of the most destitute, the new agricultural crops began to be used to produce more food for the wealthy as well. Food became a commodity to be manufactured, with the farmers providing 'inputs' to a production system, which acts as a direct feeder to Big Business, rather than 'outputs' for direct human consumption.[13]

For example, cocoa, coconut, rubber, bananas, tea and coffee were produced on 'large-scale, mono-cultural, capital intensive farms employing hired workers and with management and supervision by expatriates . . . often under control of agribusiness corporations.'[14]

Products, such as sugar and oil, which had once originated exclusively in the tropics, started being sourced closer to home through crop substitution – sugar beet and high-fructose corn syrup in place of sugar cane; soy and rape seed oil rather than palm and coconut oil. Big businesses became more interested in food production as new farming efficiencies (afforded by advances in technology) paved the way for larger and larger profits.

Regime 3 – Fordism and the Rise of 'Big Food'

Academic researchers coined the term 'Big Food' to describe the mighty conglomerates which have arisen to dominate the food markets in the US, Europe and, increasingly, the developed world.[15] The inevitable consequence of the abundance of low-cost crops, and the takeovers and mergers of the 1980s, is that much of today's heavily processed food is marketed and retailed by just a handful of huge corporations that wield tremendous economic and political power.

In the US, half of all food is produced by just ten of such corporations: Ferrero, General Mills, Grupo Bimbo, Kellogg's, Kraft Foods, Mars, Nestlé, PepsiCo, the Coca-Cola Company and Unilever.[16]

The continuing pursuit of profits in the food industry has led to the rise of what is termed 'Food Fordism'.[17] By breaking down complex manufacturing tasks into a series of simple, repetitive ones which, while requiring specialised tools, could be carried out by relatively unskilled labour, Ford created the template now used in today's food industry. The journey of food from field to factory to retailer is now organised like an industrial production line. This has enabled transnational corporations to satisfy worldwide demands for low-cost food.

Today, of the £19 billion in UK food exports, £11 billion consists of highly processed food, and £6.4 billion of lightly processed food, while only £1.4 billion is made up of unprocessed foods.[18]

The erosion of food cost and explosion of food supply has not happened without serious health consequences. It is the power of mass production and the desire to increase profit margins that has led to our supermarket shelves being filled with the cheap HED foods which are so tempting and potentially so dangerous.

The Growth of Agri-Tech

Although not among the regimes listed by Atkins and Bowler, we would add a fourth category that seems set to transform the face of the countryside for ever within the next few decades. By 2050 at the latest, farms as we know them today will have almost ceased to exist. Those few that remain are likely to be either niche producers catering to a small number of high-income consumers, or else heritage sites and visitor attractions. They will otherwise have become too costly and inefficient to survive.

Since both of us come from farming backgrounds, we make this prediction with great sadness. But, given the rise of the global population, which the UN predicts will have increased by 2.5 billion by 2050, and the need to curb greenhouse gas emission, which the UN estimates could have increased by 30% through agriculture alone in the same time, it is clear a different approach will have to be taken.[19]

Even today, 80% of the land suitable for cultivating food is already in use. By 2050 it is estimated that 70% of the population will live in cities.[20] As a result, many experts see low-energy, city-based farms as the only option to safeguard the environment whilst ensuring that sufficient food is produced. While cereals such as wheat and corn will still be grown in the open air, fruits and vegetables are far more likely to be planted and harvested by robots than traditional farmers. Livestock, too, will be tended and monitored by sophisticated, computer-controlled machinery. The people still employed in farming are as likely to be computer technicians and engineers as agriculture specialists.

This future has already arrived in the Netherlands. At Hoeve Rosa farm in the Dutch province of Limburg, robots monitor and care for the wellbeing of 180 cows.[21] They milk them when required, keep their stalls clean, and alert farmer Fons Kersten via a phone app should one of the animals need his attention. Since installing the machines in 2008, at a cost of 500,000 euros ($730,000), Fons Kersten has been able to double his number of dairy animals and has seen the yield from each cow rise between 10% and 15%. And Dutch scientists have developed robots fitted with cameras to assess when tomato plants need a shot of pesticide. These machines are reported to have reduced total pesticide usage by up to 85%.

Soon fruit and vegetables will grow not in the soil and under the sun, but in hydroponic factories with liquids providing the nutrients and LEDs the light. Located in Portage, Indiana, 50 km outside Chicago, is Green Sense Farms, the world's largest vertical farm. Here herbs and leafy greens are cultivated in 25-ft-high carousels under 7,000 LEDs. Because these lamps produce less heat than fluorescent lights, the plants can grow in much closer proximity to them. This means that rather than expending energy growing upward, the plants grow outward and become denser and leafier. Their roots are planted not in soil but in nutrient-enriched water trickled over pulverised coconut husks; not only does this keep the plants clean and free from pests, they are also reported to be ten times denser, with yields up to 30 times greater than conventionally grown plants.[22] They are also far more environmentally friendly, significantly reducing both the water use and the carbon footprint of every plant grown. 'The lettuces are more intensely flavoured and crisp,' says CEO Robert Colangelo. 'We can generate 150 heads per tub every 30 days.'[23]

The food business currently revolves around making huge quantities of energy-dense yet nutritionally poor food using the cheapest possible ingredients. This allows for excellent profit margins – but at the expense of a surge in global obesity. Problematically, the new agri-tech could actually just make matters worse by making it cheaper than ever to do that. Yet, with these new advances, fresh, non-processed foods could also be much more readily available, even in the poorest parts of the planet.

It is to be hoped that we will indeed see people in Western industrialized countries revert to more traditional ways of eating. Yet, just as we begin to reap the benefits of traditional diets once more, the crest of the high-fat, high-sweet tsunami is beginning to break on those living in developing countries. Academics and dieticians often argue that Big Food is responsible for the phenomenon known as the 'nutrition transition'. This is the change, particularly in lower- and middle-income countries, from a traditional local diet to Western-style consumption – unprocessed foods rich in nutrients and fibre are replaced by the same processed foods, high in sugar, salt and fat, with which we have become so familiar.

The Rise of Processed Food

We've repeatedly mentioned the prevalence of cheap 'processed' foods in the world today, but have yet to explore what exactly we mean by this term. It is time that we did so.

In a sense, there is nothing new about processing food. It began the moment humans discovered how to cook meat, grind grain into flour, and to pickle, salt and smoke food to preserve it. However, with the Industrial Revolution came the technology, organisation and investment capital which enabled processes such as milling, refining, distilling and fermenting on a large scale, meaning foods ranging from bread and jam to sugar, beer and wine could be mass produced.[24]

'Our ancestors ate whole grains . . . ' comments Dr Gene-Jack Wang, chair of the medical department at the US Department of Energy's Brookhaven National Laboratory, in Upton, New York, 'but we're eating white bread. American Indians ate corn; we eat corn syrup.'[25]

Unfortunately, only a minority of today's shoppers understand what constitutes a 'processed' food. They often assume that the term only refers to a small number of manufactured products. In fact, with the exception of fresh fruits and vegetables, anything that comes in a box, tin or bag can be considered processed. According to a controversial paper by the food industry-linked American Society of Nutrition, there is nothing wrong with this, since processed foods 'make up an essential part of the American diet.'

'Many staples in the diet, such as bread, cheese, and wine, bear little or no resemblance to their starting commodities,' points out lead author Connie Weaver, from the Department of Nutrition Science, Purdue University, West Lafayette. '[They] are highly processed and prepared but are often not regarded as "processed" by consumers.'[26] The logic being that, because, say, bread doesn't look anything like yeast, flour and water, it's hardly surprising that a lot of foods go through radically transformative processes. Not everyone has been entirely convinced by this line of thought.

Interestingly, the majority of members (7 of the 12) on the editorial board of the *American Journal of Clinical Nutrition*, which published the contentious article, list major corporate affiliations. 'The list of food companies for which they consult or advise is too long to reproduce,' says Marion Nestle in her blog on Food Politics, 'but it includes Coca-Cola, PepsiCo, The Sugar Association, The National Restaurant Association, ConAgra, McDonald's, Kellogg's, Mars, and many others.'[27]

While we are certainly not suggesting that anything improper occurred in the writing or publication of this article, such potential conflicts of interest could raise questions in the minds of others.

One problem we see with the article is that the authors fail to draw any distinction between the sort of 'processing' involved in washing, freezing and packing products such as fruit and vegetables, and what has been termed 'ultra-processing'.

Many ultra-processed products are created from substances extracted from whole foods, such as the remnants of animals or inexpensive ingredients such as 'refined' starches, sugars, fats and oils, preservatives, and other additives.[28] While perfectly legal, many of these products are in effect fake foods; they may be engineered to look, smell and taste like familiar foods, but they are not always quite what the consumer imagines them to be.

'At the extreme, these are foods that all but glow in the dark,' says David Katz, of Yale's Prevention Research Center, speaking of ultra-processed foods. 'On the other hand, cooking, freezing, drying, and fermenting are also forms of "processing", making grilled salmon, frozen peas, dried figs, and organic plain yogurt "processed foods" too. So much depends on just what we mean.'[29]

'Pink Slime'

One of the most controversial processed foods of all is what the meat industry describes as 'lean, finely textured beef' (LFTB). Critics have dubbed it 'pink slime' or Soylent Pink.* Used in foods such as hamburgers as an additive, LFTB is manufactured from leftover beef scraps after the production of the usual cuts of meat.

To produce it, these unwanted remnants, which might once have been used as animal feed, are first heated to around 100°F. They are then dumped into giant grinders, to deal with the fat, bone, cartilage and connective tissues, before being placed in a centrifuge. Here they are spun at 3,000 revolutions a minute while being exposed to ammonia hydroxide gas to kill bacteria. From there the meat travels to 14-ft tall drums, where it is flattened by giant rollers and frozen to produce a lighter colour. Finally, it is transferred to a grinder to produce a 60-lb 'meat brick' ready for shipment to hamburger manufacturers and other processed-food producers.

Following a controversial US TV programme from celebrity

---

* Derived from 'Soylent Green', the name of a new food in the 1973 film of the same name. Soylent Green was a government euphemism for introducing cannibalism as a way of feeding an ever-growing population. The film was based on Harry Harrison's novel *Make Room! Make Room!*

chef Jamie Oliver on the subject of this 'pink slime', many major customers cancelled their orders, forcing plant closures. The industry fought back, with Beef Products Inc., the leading producer of the filler, opening its plant in South Sioux City, Nebraska, to journalists and local governors.

'We're trying to get people to eat better and now what is going to happen because of this unmerited, unwarranted food scare, and that's what it is . . . you're going to drive up the price of lean ground beef,' Kansas governor Sam Brownback told reporter Andrew Stern.[30]

While many will find the way LFTB is manufactured unattractive, the meat is perfectly safe to eat. Amongst the vast host of different processed foods, it is arguably far from being the most dangerous from a health point of view; high-fructose corn syrup and even refined sugar are likely to present more of a health risk, given how inclined we are to overconsume sweet foods.

For this reason, we need to be equally sceptical about health scares as we are about 'super' foods. The kinds of foods that blatantly encourage passive overconsumption, such as those high in sugar, salt and fat are the ones we really need to be worried about. Other foods, however unpleasant they are to manufacture, may not be as 'unhealthy', and the sensationalistic warnings they attract derive from the way they are made, as opposed to the effect they have on our body.

As Mark Twain remarked: 'Those that respect the law and love sausage should watch neither being made.'

'Big Snack'

The phrase 'Big Snack' was coined by Carlos Monteiro from the University of São Paulo and Geoffrey Cannon from the World Public Nutrition Association in Rio de Janeiro[31] to describe companies that exclusively produce packaged, long-shelf-life products designed to replace meals.

With many of these hyper-palatable, energy-dense foods being specifically formulated to be addictive, they have become key factors in the development of obesity. The famous 1963 slogan on Lay's potato chips – 'Betcha can't just eat one' – 'might have seemed cute and innocent . . . ' comments author Michael Moss, 'but today the familiar phrase has a sinister connotation because of our growing vulnerability to convenience foods, and our growing dependence on them. "Betcha can't eat just one" starts sounding less like a light hearted dare – and more like a kind of promise. The food industry really is betting on its ability to override the natural checks that keep us from overeating.'[32]

Monteiro and Cannon share his concerns. 'Governments and international institutions now tend to cede their prime duty to protect the public interest to vast transnational corporations whose primary responsibility is to their shareholders,' they claim. 'The prevailing political, economic, and commercial policies and practices have also given these corporations freedom to expand across borders. Consequently, the leading food and drink product corporations are now colossal concerns. Their brands sell throughout the world in outlets that range from large supermarkets to filling stations, and from restaurants to kiosks.'[33]

On one point, all health professionals are in agreement. Action needs to be taken by regulators to counter the obesity-producing power of the modern snack. In Brazil, where adult obesity levels are around 14%, the caloric contribution made by such ultra-processed foods has risen from around 20% in the 1980s to 28% today and is predicted to climb as high as 60% over the next few years. It may not be too surprising to learn at this point that the nationality which consumes the greatest number of snacks, Brazilians, have a 37% higher probability of developing obesity.[34]

The Mathematics of Bliss

The food industry has invested millions of research dollars in creating snacks and other products that are, quite literally, irresistible. In his ground breaking book, *Sugar, Salt, Fat*, American investigative journalist Michael Moss describes an economic equation that captures the way our concerns for health are overridden by the pleasant and rewarding feelings associated with eating junk food. The 'bliss point' is an economic term referring to the moment where a consumer is completely satisfied; as we saw in Chapter 9, in the context of food it refers to the point where sugar, fat and salt are optimised for palatability.[35]

While we may view our own tastes as inherently unique, many human flavour preferences are about the same. For example, 10% sucrose solution achieves maximum pleasantness; and, perhaps not surprisingly, most soft drinks hover around this 10% sweetness mark, enjoying enormous popularity.[36] Sensory scientists work within the food industry to create foods that align with innate taste preferences like this.

'Humans like sweetness, but how much?' asks psychologist Robert McBride, founder of Sensometrics, a Sydney-based research company. 'Pleasure from food is not a diffuse concept, it can be measured just as the physical, chemical and nutritional factors can be measured. For all ingredients in food and drink, there is an optimum concentration at which the sensory pleasure is maximal. This optimum level is called the bliss point. The bliss point is a powerful phenomenon and dictates what we eat and drink more than we realize.'[37]

Yet, as industry veteran Howard Moskowitz told us: 'Bliss points are hard to find. We know they exist, we know they're evolved, but they also depend on the person and the time.'[38] Moreover, there is significant pressure to produce blockbuster products; 75% of all processed food concepts fail, and only 3% are capable of generating revenue for years beyond initial launch.[39]

Unread on the Label

In most regions of the world, the ingredients in food, whether heavily processed or not, have legally to be stated on the label; however, research suggests that only a minority of consumers actually take the time to read this information. This isn't all that surprising – it's typically in a small print, often further obscured through the use of coloured type.

In a study in our lab we used eye-tracking technology to analyse how much attention people pay to food labels. People were fitted with glasses that tracked the movements of their eyes, meaning we could see what parts of the packaging they looked at, in what order, and for how long. Given a packet of processed 'honey roast ham', for example, nearly all our participants looked only at the colour photograph on the front of the packaging and read the product description printed in large, easy-to-read type. Fewer than one in ten took the time to turn the pack over to study the far smaller and more difficult-to-read list of ingredients. Had they done so, it would have been apparent to them just how much additional sugar, fat and salt it contained.

Even if you do read the label, the information is often presented in complex nutritional jargon, which can mask both the number of calories and the amount of salt, sugar and fat a product contains. How many consumers, for example, would understand that 'sorbitol' and 'xylitol' are actually sugars produced by alcohol?[40]

Fast Food

For millions of consumers around the world, fast food is, over-whelmingly, the preferred way to eat out. In 1987, McDonald's had 951 restaurants worldwide. By 2002 this had grown to 7,135, and today there are an estimated 32,000.[41] In America, the number of fast-food restaurants per capita doubled between 1972 and 1999, as obesity rates climbed from 13.9% to 29.6%.[42]

Nor is fast food only available from outlets on the streets of towns and cities. More than 500 US universities and college campuses offer fast food, as do over 5,000 schools. More than 95% of American schoolchildren are able to identify Ronald McDonald, the McDonald's clown, a character whose popularity is second only to Santa Claus.[43] Unsurprisingly, due to its generally high calorie content, large (and increasing) portion sizes, and use of processed meat, refined carbohydrates, unhealthy fats, and high levels of salt and sugar, fast food has been linked with both obesity and cardio-metabolic diseases in both developed and developing nations.[44]

Despite this, the extent to which it contributes to obesity remains contentious.[45] Only recently have researchers started investigating the strength of the association between the two. The World Health Organization has found that fast-food consumption is indepen-dently and positively associated with Body Mass Index (BMI).[46] That is to say, the greater the concentration of fast-food outlets, the higher the mean BMI for the population as a whole; with each additional fast-food restaurant leading to a fractional increase in BMI.[47]

Unfortunately this also means that transnational food corpora-tions are among the leading vectors for the global spread of obesity and other non-communicable disease risks associated with an HED diet.[48] Small wonder, then, that most scientists and public health advocates support the regulation of both the fast food and Big Food industries.[49]

All Aboard the Corporate Social Responsibility Express

It was the tobacco industry that pioneered corporate techniques for influencing the public, the press and policymakers by shifting the focus of attention on their products away from the true consequences of consuming them.[50] A notorious tactic from the Big Tobacco playbook is what trans-nationals euphemistically refer to as their 'corporate social responsibility' programmes. CSRs are an 'evolving concept' that includes a company's ability to 'play nicely' and accept accountability for economic, legal, ethical and philanthropic commitments to society in addition to its financial duties and accountability to shareholders.[51] 'Cause marketing' is a type of CSR that connects a company to a specific social benefit or 'good cause', often a community initiative or organisation that benefits from the sale of a product or brand.[52] In many cases CSRs are responsible for a great deal of good work, for which their companies should be given credit. However, we should also be aware that they can be used exploitatively; giving the impression that a company is behaving responsibly when in fact this is far from being true.

By substituting an unrelated charity, the organisation circumvents the real responsibility in acknowledging the problems with their products. In the case of Big Tobacco, the cancer risk is ignored, while charitable donations are promoted and celebrated. Big Food, many critics would argue, have adopted a similar approach by encouraging healthy, active lifestyles as opposed to dealing with actual caloric value of the foods they produce. This is an indirect way of dealing with mounting concerns about the effects of their products, and the way they are marketed, on public health.[53]

The Power of Trade Associations and Lobbying

The American Beverage Association (ABA) is an industry trade group that lobbies energetically against taxes on sugar-sweetened beverages. ABA members, primarily consisting of beverage companies, are part of the front group 'Americans Against Food Taxes', which has aired a $10-million TV campaign against taxing beverages, promoting individual responsibility as the primary remedy for obesity.[54]

Encouraging personal responsibility is the first defence for both the tobacco industry and food industry. Yes, everyone should be encouraged to take personal responsibility for their health and what they consume, yet at the same time, the relationship between obesity and personal responsibility is not a straightforward one. Campaigns such as those funded by the Americans Against Food Taxes need to be considered in the context of the fact that sugary beverages can play an important role in the development of the pathologies associated with obesity discussed in Chapter 5.[55]

While in developed countries education and awareness mean that companies have to address the potentially negative consequences of consuming their products head on like this, in the developing world people are only just beginning to learn about the effects that processed food can have on bodyweight. The food industry is increasingly targeting developing countries as regions for expansion.[56] In Vietnam and India, per-capita consumption of soft drinks is projected to have doubled from 1997 to 2015, and in Egypt, China, Tunisia, Cameroon and Morocco, consumption of soft drinks is estimated to have increased by about 50% within the same time frame. The changing consumption patterns of these nations speak for them-selves: Mexicans now drink more Coca-Cola than milk.[57] Furthermore, studies have found that a 1% rise in soft drink consumption results in an additional 4.8 overweight and 2.3 obese adults per 100 of the population.[58]

The Rising Power of the Supermarkets

Between 1971 and 1995, the UK lost over 120,000 independent retailers; Spain 34,000; West Germany 115,000 and France 105,000. They had been driven out of business by the purchasing power of the supermarkets; few small grocery stores remain.[59] Today, the top five UK supermarkets sell more than 70% of all groceries bought, up from 29% in 1973.[60] In the US the largest five super-markets were found to control 48% of the market in 2005, up from 24% in 1997.[61] The number of Wal-Mart outlets alone increased by 50% between 2001 and 2005, standing at more than 11,000 today.[62]

Because these 'Big Box' retailers often purchase processed foods at large discounts, they are able to significantly reduce their price, encouraging consumers to purchase, and therefore consume, more.[63]

So, what people eat is increasingly driven by a few multinational food companies and retailers; the world's food system today is more like an oligopoly than a competitive market place of small producers. This set-up may directly impact obesity rates, as a handful of retailers buy the majority of their stock from a handful of large food companies. It may seem as though we have a dizzying array of choices in the supermarket, yet to a large extent we are simply being presented with a cascade of energy-dense foods, all more or less the same, yet packaged and reformulated to convey the idea of choice.

The Return of the Farmers' Market

There has been a certain amount of pushback against the rise of the supermarkets. The past decade has seen a rapid growth in the popularity of locally produced food and farmers' markets, the resurgence of the latter originating in California. There are now more than 8,000 farmers' markets in the US, and around 500 in the UK, while farm shops have often been transformed into extensive food halls attracting thousands of customers weekly.[64] In 2013, a survey found that a third of British households shopped in farmers' markets on a regular basis. Locally produced food, with its apparent environmental benefits and reputation for being healthier, have a powerful feel-good factor which appeals to many consumers.

Some economists, however, warn against what has been termed the 'locavore' movement.[65] There is also growing scepticism among environmentalists. While it may seem like a better way to shop, ultimately substituting farmers' markets for our current industrialised food production techniques could actually jeopardise the climate further, leading to a greater use of chemicals and increased acreage under cultivation to meet food demands.[66] In the longer term, it could also endanger world food supplies, making food scarcity for the most vulnerable an even greater concern. The high crop yields and low costs we take for granted would be under immense threat were such local markets to become mandatory or enforced by law. According to Kevin Frediani, head of sustainable land use at Devon-based Bicton College, in the southwest of England, a head of lettuce grown in Spain and then transported by road to the UK has a carbon footprint of around 1.5 kg. The same lettuce produced under glass in the UK has a footprint of about 1.8 kg. 'We just don't have the light,' he explains, 'and glass is extremely inefficient at keeping heat in.'[67]

So, despite seeming a more natural, authentic way to produce food, trying to turn back the clock to a past age of food production is not a viable solution if we are to have any hope of feeding the planet; and so we are back to the need to use efficient new agri-tech methods to mass-produce nutritious food, as described earlier in the chapter.

In summary, food is now one of the world's most heavily marketed commodities, and one of the most cleverly and astutely marketed at that.[68] And, unfortunately, the current reality is that the tastier and less nutritious a food is, the more profitable it is, too – a state of affairs which is likely to remain until new, more efficient models for growing nutritious foods are widely adopted. With their low production costs, long shelf-life and high retail value, such fat and sugar-laden foods are more likely to sell than products that may be less pleasant to eat but are significantly better for our health. Since they have a responsibility to their shareholders, companies naturally want to go for the least risky option.

Fat bottom lines indeed.

PART FIVE

LOSING IT WRONG – LOSING IT RIGHT

'Excess fat can't be blamed on insulin, carbohydrates or the Loch Ness Monster. Gaining body fat comes from taking in more calories than you burn. Anyone who can prove otherwise will surely win a Nobel Prize in physics for disproving the first law of thermodynamics. I am unaware of that particular prize being awarded.'

Losing it Right, James Fell & Margaret Leitch[1]

CHAPTER 13

Losing It Wrong:
Why Fad Diets Fail

'When rapid inane approaches to weight loss become prime
time juggernauts on television, go figure that's what people look
for in weight loss.'
Dr Yoni Freedhoff, *The Diet Fix*[1]

It was, by any standards, the ultimate fad diet.

Over a ten-week period Mark Haub, a professor of nutrition
from Kansas State University, lost 27 lb by cutting his calories from
2,600 per day to less than 1,600. As a result, his weight fell from
a little over 14 stone (201 lb) to around 12 stone (174 lb) while his
body fat decreased from around 33% to 25%. LDL ('bad' cholesterol) fell by 20% while HDL ('good' cholesterol) increased by a
similar amount. In addition, his triglyceride (a form of fat) profile
was reduced by 39%.[2]

That he lost weight by reducing the number of calories he
consumed is hardly surprising. *What* he ate while doing so created
a social media storm.

For more than two months, Professor Haub's diet consisted
largely of Twinkies, brownies, and cupcakes, with between-meal
snacks of Doritos, Corn Pops and fudge. A typical daily menu
would include: Golden Sponge Cake: (150 calories; 5 grams of fat);
Little Debbie Star Crunch (150 calories; 6 grams of fat); Doritos
Cool Ranch (75 calories; 4 grams of fat); Kellogg's Corn Pops (220
calories; 0 grams of fat); whole fat milk (150 calories; 8 grams of

fat); Brownie Chewy Fudge (270 calories; 14 grams of fat) and Little Debbie Cake (160 calories; 8 grams of fat). In addition, he ate some canned green beans, baby carrots and a few sticks of celery. He also took a multi-vitamin pill each day and drank a Muscle Milk protein shake.

To a lot of readers this probably seems like a shocking diet. However, far from regarding eating all this junk food as unusual, Haub claimed it was an accurate reflection of what millions of people eat every day. Furthermore, he thinks 'it's unrealistic to expect people to totally drop these foods for vegetables and fruits. It may be healthy, but not realistic.'[3]

We must emphasise that neither we nor Professor Haub are recommending such a diet as a healthy way to lose weight! The long-term consequences of avoiding fruit and vegetables are likely to prove serious. However, what this unusual experiment demonstrates is that when it comes to losing weight, it's calorie-cutting which matters most. Provided the number of calories you consume remains low, you could just as easily slim by eating pork pies or hamburgers than nibbling on lettuce leaves or nuts. You would just have to ensure that you were eating smaller portions in the case of high-calorie foods. Equally, as we saw in Chapter 9, it is possible to put on weight by eating foods widely considered to be healthy, such as oranges; all you need do is consume enough of them.

Recent years have seen the rise of 'fad diets', weight-loss plans which promise dramatic results through beginning a very specific (and usually very limited) diet. They may include some kind of specific 'miracle' ingredient, the dieter being told to eat one particular foodstuff in large quantities. They also frequently involve the elimination of a particular macronutrient from the diet, almost always either fat or carbohydrate.

The two differ in energy content, with fat providing the greatest amount, at 9 calories per gram, and carbohydrate approximately

4 calories per gram (which is the same amount as protein).[4] The pendulum of fashion swings back and forth between low-carbohydrate versus low-fat diets, with one being favoured, then the other. Each side of the debate has a slew of academic and dietetic researchers proclaiming its health benefits. So, let's consider the role each plays in our diet to better understand why either one might be considered a candidate for elimination.

Eliminating Fat

During the 1990s, many diets advocated reducing or banning fat due to its high energy content. With what is in fact faulty logic, we believed that eating fat would inevitably make you fat. This is far from the truth – weight gain results from consuming more calories than your body requires, irrespective of whether the food is fatty or not. Yes, eating enough calories in the form of fat will result in weight gain, but the same is true for protein or carbohydrate. Furthermore, at least 20% of our diet *must* consist of fat, since, without it, fat-soluble vitamins (A, D, E and K) cannot be transported, stored or absorbed.[5] So while it tends to get something of a bad press, it is actually essential to have fat as part of a healthy diet. That said, different kinds of fats have different benefits and dangers, which it is worth being aware of.

Saturated fat

Typically derived from animal sources such as red meat, poultry and full-fat dairy products, saturated fat raises total blood cholesterol levels and low-density lipoprotein ('LDL'), raising your risk of developing cardiovascular disease if consumed in excess. Overconsumption of saturated fat has been associated with the development of Type II diabetes.

Trans fat

Although trans fat occurs naturally in some foods, most trans fat occurs in oils that have been partially hydrogenated. Hydrogenation is the process that makes the fat easier to cook with, and also less likely to go bad. This makes it an ideal ingredient in the packaged goods segment, and it has been used in everything from crisps to cakes. However, research has shown trans-fat consumption to be associated with an increase in the risk of developing heart disease, diabetes and perhaps of strokes. It is therefore advisable to consume as little of this type of fat as possible.

Trans fat has been the focus of numerous health campaigns, and the food industry has responded by eliminating trans fats in many of their products; thanks to significant pressure on behalf of the public health community, it is now becoming an industry standard to refrain from using trans fats in foods.

Monounsaturated fats

Monounsaturated fats are liquid, but can turn solid when chilled. They have been shown to decrease risk of heart disease, and may benefit insulin levels and blood-sugar control. Olive oil is one of the most famous monounsaturated fats; other examples include avocado, almonds (raw), peanut butter, macadamia nuts, sunflower oil and seeds, and hazelnuts.

Polyunsaturated fat

Polyunsaturated fats are found primarily in plant-based foods and in oils. They occur in fatty fish such as salmon, mackerel, herring and trout, and also in walnuts and sunflower seeds. They have been shown to improve blood cholesterol levels and reduce the risk of heart disease.

Omega-3 fatty acids

Omega-3 fatty acid plays an essential role in health and must be obtained from our diet, since the body cannot manufacture it. There are three types of omega-3 fatty acid. The first is called alpha-linoleic acid (ALA), the second eicosapentaenoic acid (EPA) and third docosahexaenoic acid (DHA). Fish are the best source of omega-3s; although they can also be obtained from plants, the body does not metabolise these as effectively as it does those from fish sources. Omega-3s help protect against heart disease, control blood clotting and build cell membranes in the brain.

Fad diets that are high in fat fail to take into account the fact that such a diet can wreak havoc on blood lipid levels. It is certainly not the 'healthiest' way to eat.[6] However, without fats essential functions of the body such as vitamin absorption and protein synthesis cannot be carried out. So while we should certainly be cautious about the amount and kind of fat we eat, we should also realise that fat is not a foe. As with any other element of our diet, the old saying obtains – everything in moderation.

Eliminating Carbohydrates

The other food group which fad diets frequently focus upon is carbohydrates. Low-carbohydrate plans can prove wonderfully effective in the short term (i.e. the first six months), but whether they are the best path to maintaining a healthy weight in the long term is questionable.

Carbohydrates can be divided into two broad categories. The first group comprises the simple sugars, glucose, fructose, sucrose (glucose + fructose) and lactose (glucose + lactose).[7] These are found in foods such as flour, bread, pasta, some vegetables such as potatoes, and fruits. Once consumed, they enter the bloodstream via the small intestine and travel to the liver, where they are converted into glucose.[8] Excess glucose is stored in the liver and skeletal muscles as glycogen. Should the stores become full as a result of overconsumption – they can only hold about 2,000 calories – glycogen will then be stored as fat, resulting in weight gain.

If, on the other hand, there is insufficient glucose in the bloodstream, the body responds by breaking down fat. However, after an extended period of time, the body will begin to rely on breaking down amino acids. By doing so it prevents protein (where amino acids are found) from serving its main purpose of building and maintaining muscle. It also places additional stress on the kidneys as they are obliged to remove excess protein from the blood.[9]

The second group is the complex carbohydrates: starch and fibre. There are three key components to grains: bran, germ, and endosperm. The bran is the outer shell of the grain, and provides fibre and B vitamins. The germ contains E vitamins and essential fatty acids. Finally, the endosperm contains starch. For this reason, consumption of whole grains (i.e. those that keep the bran and germ) is encouraged as a way to increase consumption of micronutrients and fatty acids.

Fibre comes in two forms, insoluble and soluble. Insoluble fibre is found in the outer coat of whole grains, the stringy bits of celery and the skins of fruit. Soluble fibre is what gives vegetables, fruit, grains and beans a thicker, sometimes gummier texture. It is also present in oatmeal, apple sauce, prunes, beans and lentils. Insoluble fibre regulates gut motility, while soluble fibre lowers blood cholesterol levels. Critically, both types contribute to feelings of fullness after a meal and, because they take significantly longer to digest than simple sugars, offer a more sustained source of energy.

However, as is the case with simple sugars, complex carbohydrates are also converted into glycogen, and if not used by the body are stored as fat. So, the 'low-fat' and 'no-fat' diets of the 1990s encouraged people to consume grains, bagels and gummy bears to their hearts' (dis)content. Then, when researchers reported that the insulin-spiking effects of sugar can promote weight gain and fat storage, this type of diet suddenly became significantly less attractive. Moreover, as we learn more about the rewarding effects of the taste of sweetness, and how this is processed in the brain, we are becoming increasingly wary of sugar and related sweetening substances.

So again, a balance needs to be struck: consuming too many carbohydrates of either type will lead to weight gain and fat storage, but failing to consume enough will prevent the body from functioning healthily.

The Frankengrain?

Wheat is a source of both simple sugars and complex carbohydrates and, for many people the world over, wheat-based products are a staple energy source. However, in recent years, two *New York Times* bestsellers have been devoted to exploiting their supposed evils. *Wheat Belly*, by cardiologist William Davis, and *Grain Brain*, by neurologist David Perlmutter, encourage us to believe that a multitude of health problems are due to grain intolerances and allergies.[10] But is this really the case?

Davis describes wheat as the 'perfect Frankengrain', and accuses it of inflicting more harm than 'any foreign terrorist group'.[11] The sensationalist nature of this assertion aside, there is no doubt that there are some drawbacks to eating wheat. Coeliac disease, a serious medical condition, is relatively widespread and is greatly aggravated by wheat products. A coeliac suffers from an autoimmune reaction to gluten (a protein found in wheat, rye and barley) in their small intestine. This interferes with the absorption of nutrients from food. For a coeliac, even a mouthful of a food containing gluten can produce considerable discomfort, abdominal pain, gas, bloating and constipation. Gluten is not only found in food but also in medicines, vitamins and even lip balms. Recent research also suggests that coeliac disease can lead to far more serious conditions, including anaemia, osteoporosis, sterility, and possibly even cancer.[12]

Leaving such medical problems aside, it is true that skyrocketing levels of obesity are likely linked to significant increases in the consumption of processed carbohydrates, including wheat. However, in spite of the negative press it has begun to receive, it is hard to see why wheat should be singled out for particular criticism.

Many of the cereals and foods we eat today are hybrids, the result of mixing between different strains. This is sometimes cited as the source of our problems with wheat and corn. Yet the same agricultural process has been carried out for tens of thousands of years without producing today's skyrocketing levels of obesity. Furthermore, while some people do suffer from food intolerance and food allergies, doctors estimate these afflict only around 1% of the population.[13]

It seems likely that a significant proportion of those claiming to be 'gluten intolerant' are actually experiencing what doctors refer to as 'gluten avoidance'. Gluten avoiders have stomach problems without knowing why they occur or how best to address the underlying causes. Rather than seeking medical advice, or perhaps being unconvinced by the solutions their physicians offer, they search the Internet and learn about the many so-called 'benefits' of going gluten free, rapid weight loss being amongst them.[14] This can indeed occur, although most nutritionists suggest the most likely reason for this is a placebo effect – in essence the person modifying their eating behaviour and consuming less *because* they are following certain dietary restrictions, rather than the particular types of food they are allowed being an effective weight loss regimen in themselves. This is a fairly common occurrence when an individual is embarking on a new diet.[15] In recent years gluten-free products have become a food marketer's dream come true, with a US market value of $4.2 billion. The sector is also projected to grow by 10% until 2018.[16] The range of gluten-free products now extends from pasta to church communion wafers! Even Kellogg's has jumped on the bandwagon, removing barley malt in order to advertise their classic Rice Krispies are 'easy for kids to digest'.[17] Gluten, and by extension wheat and other cereals that deliver protein (indeed, gluten is a protein), has been blamed for everything from excess tummy fat to mood swings. However, many gluten-free products use rice and tapioca flours, which have little or no nutritional value. They often sit high on the glycaemic index, in addition to being higher in fat and lower in protein, which makes them less satiating.

'We have a tendency to think that gluten-free is healthier,' says Toronto-based naturopathic doctor Meghan Walker, but she asserts that, contrary to our expectations, 'that is certainly not the case.'[18] In fact, research shows that adopting a gluten-free diet can result in higher weight gains than following a diet high in protein but low in processed carbohydrate. The unfortunate fact about many gluten-free substitutes is that they are higher in calories – and fat – than the gluten-filled original. A stamp of 'gluten free', just like its 'fat free' cousin of the past, does not mean a licence to binge.

Carbohydrates: the Good, the Bad and the Ugly

Just as certain types of fat should be avoided, the same can be said for certain kinds of carbohydrate. Processed carbohydrates and flours stripped of their whole grain shell (white flour), for example, are excellent at delivering calories but offer very little else. White bread is devoid of the health benefits of whole grain because, with the removal of the outer shell from the grain, the insoluble fibre it would otherwise have contained has been lost. Moreover, white flours are also digested quickly, raise blood sugar levels quickly (and so also cause them to drop quickly), and therefore do not have the same satiating benefits as wholegrain options.

More than any other form of carbohydrate, High Fructose Corn Syrup (HFCS) has been blamed for the obesity pandemic. This comes in two forms: HFCS 42, which consists of 42% glucose and has the sweetness of table sugar (sucrose), and the even sweeter HFCS 55, with about 55% fructose, which is mainly used in soft drinks.[19] HFCS is produced by refining corn kernels to a point where long chains of glucose are broken down into short chains of fructose. Due to the Farm Bill in America, corn is cheaper than cane sugar in America. This means that many US-manufactured products contain HFCS – the government is in effect encouraging production and consumption of a foodstuff we know contributes to overeating, due to its ultra-sweet taste. However, there is no hidden ingredient in corn that automatically precedes morbid obesity, so a fear of corn itself isn't really warranted. As we have seen time and again, there are no short-cut answers in the obesity pandemic and, while HFCS and other sugars are undoubtedly a concern, things are not that clear cut.

'It's hard to prove that sugars are bad at all because high sugar intakes tend to cluster with other health behaviours,' explains nutritionist Laura Forbes.[20] 'There are links between corn consumption with obesity and diabetes, in particular, but we are far from having proof of causality. The main point is that high-sugar foods are low in micronutrients and vitamins, have low satiety factors, and ergo may be involved in causing obesity and diabetes. An important rule of thumb is to avoid ultra-sweet foods in general; given their propensity for causing binge eating, it may be best to put away the sweet stuff indefinitely.'[21]

Protein Power

While the body must expend between 5% and 15% of the energy contained in carbohydrate in order to digest and use it, this figure jumps to between 20% and 35% when it comes to processing protein.[22] Furthermore, while the body can store excess carbohydrate energy as glycogen (and excess glycogen as fat) it cannot store excess protein; it has to be processed right away, burning up some energy along the way in a process called thermogenesis.[23] So, proteins take more energy for our body to digest and make use of, meaning that caloric intake in the form of protein will actually result in less energy being made available to the body than if the same number of calories was ingested as carbohydrate. And since protein can't be stored as energy, there is less risk of excess consumption leading to weight gain.

When carbohydrates are largely eliminated from the diet, as described above, the body has to use fat as a source of energy in order to metabolise protein, a process called ketosis. This results in the production of ketones, which further dull the appetite.

On the basis of all this, removing most carbohydrates and significantly increasing the amount of protein we consume sounds like a good idea. Indeed, this is the basis of the famous Atkin's Diet. But is such an approach really a 'magic bullet' resulting in rapid and lasting weight loss?

The answer is both yes and no.

Yes, because such an approach is likely to work in the short term and does indeed cause fat to be consumed. No, because it offers little hope of sustainable weight loss and may compromise our long-term health.[24]

'It was once thought that the high protein intakes would help preserve lean body mass, such as muscle, when weight loss happens,' explains Laura Forbes, 'but research has shown that not to be the case – lean body mass is still lost in these diets. We also know that in people with poor kidney function, high protein intakes could increase the risk of kidney disease. At present this is more of a hypothesis than something that has solid proof. There's really no information about its long-term safety or health effects.'[25]

Thus, it is likely that high protein diets may only work for the first six months of a regime. After that point it may be advisable to slowly reintroduce complex carbohydrates in order to avoid lean body mass being shed along with fat. Since high protein will also put additional stress on renal function and the kidneys, it merits consideration whether this is really a healthy lifestyle to pursue long term.

No Matter What You Eat – It's the Calories That Count

So, despite people's belief that low-fat or low-carbohydrate diets are the answer, in fact neither one is a sure-fire route to maintaining a slim body in the long term, and both can have negative impacts upon health. Maintaining a balanced diet that incorporates all the macronutrients is far preferable.

So if there are no short cuts through eating particular kinds of food, we are forced to return to the point made at the beginning of this chapter: that the most important thing one can do to achieve weight reduction is to limit the total amount of calories one consumes. This may not be popular, exciting or new information and therefore probably doesn't have the appeal of a fad diet, but it is nonetheless true.

The average man requires around 2,500 calories daily and the average woman 2,000. The problem, of course, is that none of us is average! Our actual daily requirements vary according to factors such as weight, age, sex and level of activity. Marines, for example, can eat 5,000 calories a day or more without putting on a pound of additional weight. But then very few people exercise as hard as a marine!

Your basic daily calorie requirement can be calculated using a formula developed by scientist Max Kleiber in 1932:

Calories Required = $38W^{3/4}$ where W = your weight in pounds

If the prospect of working through an equation like that leaves you cold, you can easily find out how many calories you consume each day from our Fat Planet website (www.thefatplanet.com). On the other hand, if you are not put off by the idea of some basic maths then read on.

To calculate a ¾ power of your weight (W), multiply it (in pounds) by itself three times i.e. W x W x W. Now take the square root of that number, *twice*. (On a calculator, just hit the √ button twice). That has given us $1W^{3/4}$.

Finally, multiply the result by 38, bringing the total to the $38W^{3/4}$ of Kleiber's equation. The number on the screen will then be your total recommended calorie count for one day.

Let's run through an example.

A person weighing 150 lb would multiply this number (W) by itself three times (i.e. raise to the third power) = 150 x 150 x 150 = 3,375,000.

Now take the square root twice. First time = 1,837; Second time = 42.86. Finally multiply this number by 38 (42.86 x 38) = 1,629 calories per day.

Note that this is a *basic minimum* – it is considerably lower than the average of 2,000 calories for a woman and 2,500 for a man. If you are active, exercise on a regular basis or do hard physical work then your requirements will clearly increase, potentially by a significant amount.

Fad diets represent sensationalised adaptations of scientific theory, which fail to take into account the complexity of our relationship with food. The result, more often than not, is the development of a binge-purge cycle in which the pounds shed are quickly replaced when the dieter gives up and returns to their normal pattern of eating. Then, unhappy with having regained the weight, they begin the fad diet once more – or possibly start a new one – and the process repeats itself.

'I think people are tempted by crash diets in part because it's in our nature as a species to want things to happen quickly,' says Dr Yoni Freedhoff, Canada's pre-eminent obesity expert and bestselling author of *The Diet Fix*. 'In part because that's what the media and entertainment industries have normalised as the way to do things.'[26] We will explain why this is so in the next chapter, where we examine the traps that lie in wait to snare the unwary along the road to sustainable weight loss.

CHAPTER 14

Losing It Right:
Achieving Sustainable Weight Loss

'The first principle is that you must not fool yourself – and you
are the easiest person to fool.'
Richard Feynman[1]

When it comes to losing weight and ensuring it stays lost, there
is good news and bad news.

In the two previous chapters we focused on the bad news, drawing
attention to the fact that while restricting or removing macronutrients leads to weight loss, this weight rarely stays off for more
than a few weeks, which can lead to an unhealthy binge/purge cycle.

Now for the good news! Losing weight and sustaining that loss
is achievable, provided some lifestyle changes are made. These
involve the kind of strategies mentioned in previous chapters, such
as learning how to control stress and manage negative emotions,
having at least seven hours' sleep each night and taking regular
exercise, as well as regulating both what is eaten and how much.
In this chapter we will consider these last two factors since, in our
experience, they present the greatest challenges.

Choosing the Foods to Eat

It's something of a common-sense statement, but the key here
is to avoid foods that, while rich in calories, are poor in nutrients.

By choosing instead foodstuffs that are nourishing and which also enhance feelings of fullness (See Table 1), the likelihood of persisting with a health-promoting new eating plan is greatly increased. Research has shown that diets high in lean protein, fibre and complex carbohydrates have the greatest success in ensuring sustainable weight loss.[2] Ingredients that fulfil these criteria include chicken, fish, soya, nuts, green vegetables, fruits and whole grains such as porridge.

While there can be very little room for cakes, biscuits and sweets in a diet designed to reduce weight and enhance lean body mass, they should not be entirely excluded. Banishing every such food from your diet whilst simultaneously reducing the number of calories consumed increases the risk of succumbing to an eating binge. Food deprivation is often a precursor to binge eating as it makes that food more rewarding upon consumption.

Table 1: Choosing the right diet strategy

Food strategy	Important because
Include protein: poultry, soya, tofu, fish, nuts	Increases thermogenic effect (you burn more calories, because protein requires more energy to metabolise); it also helps you feel fuller after eating, reducing your desire to 'top up' on high energy-dense snacks
Include fruits and vegetables ('low energy-density')	Leave the stomach more slowly, promoting satiety and releasing energy more slowly
Include high-fibre foods such as whole grains, fruit, vegetables	Not so energy-dense Require more chewing Slow digestion and gastric emptying Delay glucose absorption

Food strategy	Important because
Include low-fat foods	Lower energy-density
Include foods low on the glycaemic index	Delay glucose absorption
	Improved profile for gut and satiety peptides
Minimise alcohol consumption	Increases self-control over food consumption
	Affects appetite hormones, and gut hormones
Increase consumption of polyunsaturated fatty acids	Increases thermogenic effect
Include high-calcium foods	Improve energy metabolism
Include pre- and probiotics	Improve satiety and gut peptides
Include spices, caffeine, catechins (green tea)	Increases thermogenic effect

Choosing How Much to Eat

'Don't overeat' may sound like a simple enough instruction, but in fact even the most determined dieters can be nudged off course and eat more than they intend. The reasons this happens can be very subtle; here are ten ways in which overeating can occur, together with practical strategies for avoiding this.

1: Portion control
When it comes to the amount of food we serve ourselves, perception is everything. If we are using a large plate it takes more food to make

us decide that we have a suitable portion than if we are using a small plate. This issue of size applies to both the crockery the food is served from and the utensils it's served with – the larger they are, the more we tend to serve ourselves.[3] Even if it's only the serving dish that is bigger, the portion of food served from it can increase by as much as 57%.[4] But to flip this around into a positive, a smaller portion on a smaller plate will look as satisfying to us as a larger quantity on a bigger plate. Therefore monitoring the size of the crockery and utensils you use can really help to reduce overeating.

The same is true of how we serve ourselves drinks. When asked to pour an exact measure, without the use of a measuring device, even professional bar staff have been found to pour up to 30% more into wide glasses than into thin ones. This is known as the horizontal–vertical illusion; we focus on the height of the liquid in the glass while ignoring the width, thus deceiving ourselves about the true amount. In a study we conducted, those given a wide tumbler poured almost twice as much as those with a narrow glass (397 ml vs. 796 ml).[5] Giant bottles of cola have been shown to increase the amount people serve and consume by up to 45%.[6] The lesson to be learnt from this is that using a tall, narrow glass rather than a short wide one will minimize the number of liquid calories consumed.

A lack of colour contrast between the plate and the food can also make a substantial difference to the amount served. When the two are similar, larger portions are usually taken, since it becomes harder to gauge the amount of food on the plate. We put this to the test in an experiment where participants self-served pasta in tomato sauce onto either white or red plates. When pasta in tomato sauce was dished up onto red plates, 22% more food was taken than when white plates were used.

Weight-reducing strategies:
- Avoid buying large value packs if you can afford to do so. While they may help your budget they could result in over-

eating. If you do purchase them, separate the contents into several smaller containers, then serve yourself from one of these.

- Don't cook large amounts of food, intending to eat it over the next few days – you're more likely to serve more than you need. Cook an appropriate amount for the number of people who'll be eating.
- Get rid of your large serving spoons and large dinner plates; they'll only encourage overeating. Never eat ice cream from the tub, and avoid eating sweet foods (including sugar-sweetened cereals) with large spoons.
- Consider the contrast between food and plate. Although the difference in portion size when both are similar in colour is only likely to be modest, even a small reduction in consumption can have a big effect over time.
- Drink from taller, thinner glasses rather than shorter wider ones.

2: Leave evidence of eating

One of our messiest pieces of research involved hundreds of chicken wings in barbecue sauce. While a tasty food when hot and fresh, they quickly become sticky and thoroughly unappetising when cold and partly eaten. Our purpose was to investigate whether having evidence of the number of chicken wings already consumed left on the plate would result in people eating less of them.

The participants in this study were an audience attending a comedy show in a London pub. They were provided with an unlimited supply of barbecue chicken wings to munch. Before the show began, the audience had been divided into two groups (though they were unaware of this). One group had their leftover chicken wings left on the plates as they were served with more. The other had their dirty plates swiftly replaced with clean ones by the serving staff. The group that could see the chicken bones piling up on their plates consumed an average of seven each, while those whose dirty plates were replaced averaged eleven wings per person.

A chicken wing with the skin contains 62 calories and the barbeque sauce adds another 20 calories, making 82 calories per wing. This meant that every member of the 'clean plate' group ate just over 900 calories, while those who kept their leftovers consumed 574 calories.

Similar findings have been reported by other researchers across a variety of situations; without some kind of reminder, most people struggle to remember how much they have consumed.[7] One study found that only five minutes after eating a meal at an Italian restaurant, around a third of diners (31%) were unable to recall how much bread they had eaten.[8]

By keeping the evidence before your eyes you can constantly monitor how much you have consumed and avoid mindlessly overeating.

Weight-reducing strategies

- At home, avoid clearing the table until the meal is eaten.
- Leave sweet and chocolate wrappers in plain sight rather than tidying them away immediately; it may help you think twice before going back for more.
- Avoid buying bulk sizes of treats, since this increases consumption. There is a tendency to finish an entire jumbo-sized bag of snacks at a single sitting.
- When buying treats, choose things that can be dipped in and out of and so made to last all day, such as M&Ms or Smarties, rather than things that need to be consumed right away, such as bars of chocolate. But still stick to standard-sized packets!

3: Don't be misled by low-fat labels

In a study we developed for the television programme *Secret Eaters*, amateur artists were invited to a life-drawing class, on the pretext that we were filming a documentary about painting. During a break they were offered three types of cake as an afternoon snack: banana cake, carrot cake and brownies. Two separate serving tables were set up and the artists were randomly directed to one of them. While the same cakes were available on both tables, they were described in very different ways by signs on each one.

The cakes on one table were said to be lower in fat, with an accompanying photograph depicting images of health. On the second table the cakes were described as indulgent and 'devilishly' good, with pictures implying they were delicious but not very good for you.

The number of cakes taken from each table was counted and any uneaten cake portions carefully weighed. The results were striking: those eating what they believed to be the 'low-fat' variety consumed just over 1,000 calories more on average than those offered the 'nice but naughty' option.

The danger here is obvious; the perceived health halo of a 'safe' low-fat food can tempt consumers into guilt-free overconsumption. Shoppers often take such claims at face value without bothering to check, or even consider, the extent to which such products can really be described as low fat, or the fact that even a product which really is low fat can cause you to put on weight if eaten in sufficient quantity.

When we conducted a study into how people read labels, using eye-tracking technology, we found the majority only looked at the caloric information, not the fat content or macronutrient distribution, even though it was usually printed in big, bold, easily read text on the front of the packet, bag or bottle.

We're all likely to eat more of a food we believe to be 'low fat' or 'reduced fat', and to significantly overestimate the extent of the reduction involved. While reduced-fat foods are, on average, around 11% lower in calories, many consumers believe them to be up to 40% lower.[9] This in turn can lead them to consuming up to 50% more food at a sitting.

'For normal-weight people, low-fat labelling increases consumption the most, particularly with foods that are believed to be relatively healthy,' say Brian Wansink, an established expert in eating behaviour from Cornell University, and Pierre Chandon from the world-renowned INSEAD institute in Paris; 'for overweight people, low-fat labelling increases their consumption of all foods.'[10]

Also, while we may view hamburgers to be less healthy than a sandwich containing fresh vegetables, this is a somewhat oversimplified view. For example, an advertisement depicting a foot-long Subway Sweet Onion Chicken Teriyaki sandwich claimed that it contained 'only 10 grams of fat' and compared it to a Big Mac which 'contains 33 grams of fat'. The advertising copy quoted Jared Fogle, a Subway spokesperson, as saying that this meant: 'You can eat another and another over the course of three different meals and still not equal the fat content of one Big Mac.'

He's absolutely right.

'Yet,' as Wansink and Chandon point out, 'the advertisement fails to mention that the . . . foot-long sub contains more calories (740 versus 600 calories) and more cholesterol (100 versus 85 milligrams) than the Big Mac. Thus, eating three Subway subs would provide 1620 more calories and 215 more milligrams of cholesterol than eating one Big Mac.'[11] Suddenly it doesn't sound like such a healthy option.

Weight-reducing strategies

- Be cautious about any food described as being especially healthy or in some way 'good' for you. 'Low fat', 'reduced

calorie', 'low carb,' 'natural', 'high protein', 'vitamin forti-
fied', and ' high fibre' are all phrases that can lead you to feel
comfortable eating more of a food than you would otherwise.

- Stick with the regular versions of your favourite products but
eat less of them. Not only do full-fat versions generally taste
better, but low-fat versions are often full of additives to
compensate for the lack of fat. Overall, low-fat foods may
have less flavour, so you'll overcompensate by consuming
more.

- If you do go for a low-fat product, make sure you know *how*
reduced the calorie content is, and don't overcompensate by
eating more.

- Focus on eliminating foods that combine fat with sugar, as
the combination between these two macronutrients is ultra-
pleasant, and can easily elicit overeating.

- Eat protein before any other macronutrient to help enhance
satiety. Fat-rich foods should be avoided, but if they are part
of a meal then eat them last of all when you already begin to
feel satiated.

4: Introduce pause points

A pause point is simply something that obliges you to stop eating
for a moment, during which time you can reflect on whether you
have eaten enough. When snacking, for example, a pause point
might be unwrapping a sweet, opening a new tube of Pringles or
reaching for another bag of nuts. During a meal it might be created
by having to get up from the table and go into another room, or
to a different place in the same room, to serve yourself with further
food.

In one study, every seventh Pringle in the tube offered to participants was dyed bright red to make them stop and think when they
reached it. Under these conditions, 48% fewer Pringles were eaten
in a single sitting, saving 250 calories.[12]

And in a study we conducted ourselves, we provided two groups of medical students taking part in a netball contest with bags of crisps as a mid-game snack. While they all received an identical amount, one group were given their crisps in a single, jumbo-sized bag while the others were given them in six smaller bags. Based on previous studies, we hypothesised that players given the jumbo bag would consume significantly more crisps than those obliged to open a fresh bag each time they finished one. This was indeed what happened. Students with the jumbo bag ate around 60% more than those given the separate ones.

Pause points are just as helpful when it comes to main meals. Research has shown that diners are likely to eat 20% *more* if the serving dish is left on the table while the meal is eaten.[13] (See also 'Make eating more of an effort'.)

To put this to the test, during filming of an episode of *Secret Eaters*, two groups of friends were offered a free roast chicken lunch at a fashionable London restaurant. One group ate with the chicken, potatoes and other types of vegetables in large serving dishes in the centre of the table. The second group enjoyed the same menu but served themselves from a counter in a different room. This 'pause point' meant they ate up to 50% less than those who merely had to reach across the table and help themselves to whatever they fancied.

Weight-reducing strategies

- Visual clues and pause points can be helpful reminders that you may have already eaten enough.
- Choose snacks that are individually wrapped, and so have their own built-in pause points.
- Ensure you know exactly how many calories are in each unit of a food you consume. A single almond, for example, has 7 calories. On the other hand, there are 9 calories in cashews, which provide four times less fibre than almonds. Not only

are almonds less energy-dense but they are also satiating for longer.

- Measure out snacks into little portions and eat from a bowl not from the packet.
- When serving a meal, plate up away from the table, so you can't access seconds quickly.
- If you're at a buffet or canteen, sit with your back to the food and as far away from it as possible.

5: Make eating more of an effort

The 'free chicken lunch' experiment mentioned in the previous section also illustrates what is termed the Principle of Least Resistance, first proposed by the French philosopher Guillaume Ferrerò in 1894 and later developed by the American psychologist Edward Chace Tolmen during the 20th century.[14]

Simply put, it states that people prefer to do short and simple tasks rather than lengthy or difficult ones. While this might appear to be little more than common sense, its implications when it comes to weight control are considerable.

In 2011, food psychologist Paul Rozin and his team from the University of Pennsylvania explored the subtle factors that lead to overconsumption in a cafeteria setting. The focus of their investigation was the university's refectory, which served around a thousand people each day. Their aim was to discover if diners could be influenced to eat more or less without realising this was being done.

In the experiment, eight ingredients – broccoli, grated cheese, chicken, cucumber, hard-boiled eggs, mushrooms, olives and tomatoes – were placed in various different locations on the self-service counters. In one configuration, a particular item was placed on a single large tray in the middle row of the serving counter. This meant it took very slightly longer to reach that food than those at either end. By switching around the location of high-calorie (e.g. eggs, chicken, cheese and olives) and low-calorie items (e.g. broccoli, cucumbers and tomatoes), the team found they were able to covertly manipulate diners' food choices, and therefore the number of calories consumed. For someone eating at the cafeteria five days a week during term time, the difference amounted to 3,527 calories more or less over the course of a year.

'This would translate into a 1.01 pound (0.46 kg) body weight difference,' comments Paul Rozin. 'Bearing in mind that the average annual weight gain for an adult American is in the range of 0.9 to 2.5 pounds a year, these manipulations would diminish weight gain to an extent that could have an impact on public health.'[15]

When it comes to snacks, keep them hidden from view and harder to access immediately. A study we conducted found that when a bowl of chocolates is placed on an office worker's desk, he or she will eat around nine a day; when they are placed in a desk drawer the number drops to six; and when located in sight but placed six feet away, to just four.

Weight-reducing strategies
- Biscuits, cakes, chocolates and similar HED foods should be stored out of sight rather than left in open view.
- Restrict food on display to fruit only.
- At work, snacks should never be left on desktops or within easy reach.
- When having a sit-down meal, leave the serving dishes in the kitchen, or at some distance from the table, so that you have to leave your seat to have seconds rather than just reaching across the table.

6: Avoid night-time snacking
Night-time eating poses a problem for many people who struggle with their weight. Part of the issue with eating at night is that

'satiety responsiveness', which describes the body's ability to detect hunger, is lower at night than during the day. This significantly increases the risk of overeating.[16]

Some people, perhaps aware of this problem from personal experience, resort to the habit of 'grazing' throughout the day in the hope this will reduce after-dark hunger. It is an approach to eating that, until recently, many dieticians and food writers actively promoted. Unfortunately, because this type of eating can confuse ghrelin signalling (See Chapter 6), which is essential in diminishing the desire to eat, it tends to increase calorie consumption. For this reason, the shift to eating three square meals a day is becoming fashionable again.[17]

Weight-reducing strategies

- Start each day with a healthy breakfast (see 'Eat more protein – especially at breakfast', below) containing close to 500 calories. Any less won't count as a meal, any more might make you drowsy.
- Do not eat outside of meals. This sounds simple, but is surprisingly challenging. Spending an entire week without a single snack will likely be too ambitious at first. Start by trying to avoid any kind of snacking just two days per week. Gradually increase your lack of snack to three days, then four days, and so on – until you can do this seven days a week. Try to rotate one day without snacks and the next with snacks during the working week. Choose Monday and Friday as snacking days initially since both are associated with emotional lows and highs, which may elicit the desire to eat.
- If you can't spend a whole day without snacking at first, take your lunch meal and split it into two. Eat the first portion for lunch and save the second as an afternoon snack.
- Keep a diary of how often you snack to get a better idea of your eating behaviour. A precise record isn't necessary; just keep a listed tally of everything eaten outside of meals.

● Go to bed an hour earlier. You'd be surprised – this can work wonders!

7: Eat more protein – especially at breakfast

In her 1950s book, *Let's Eat Right to Keep Fit*, American nutritionist Adelle Davis advised readers to 'eat breakfast like a king, lunch like a prince and dine like a pauper.'[18] Half a century on and this advice still holds good, even though these days most people ignore it. They do so for a variety of reasons, the two most common being they 'aren't hungry' or they 'don't have the time'.

This is a mistake that those seeking to control their weight should avoid. Because skipping breakfast increases the desire to snack, it is strongly associated with overeating, weight gain and obesity. Adolescents who skip breakfast typically snack on more desserts, high-fat salty foods and fizzy drinks compared to those who consume breakfast.[19]

Researchers have found that doubling the protein content of your diet to between 20% and 30%, from the normal 10% to 15%, can result in an immediate loss of body weight.[20] It also helps to control calorie intake over the remainder of the day in four key ways:

1. By increasing feelings of fullness for longer, it greatly reduces the desire to snack.
2. By preventing the mid-morning energy dip described in Chapter 10, by 'smoothing out' the release of energy from carbohydrates.
3. By increasing metabolism. Because the body cannot store protein, it has to process it metabolically, which uses energy and so consumes calories.
4. By reducing the desire to eat late at night by improving sleep (see 'Avoid night-time snacking' above), also by raising metabolic rate while sleeping.[21]

Increasing protein may be the first step in a series of changes that lead to achieving a healthy body weight. This is not to say that we advocate the kind of fad diets which emphasise protein to the exclusion of many other foods, but taking advantage of its satiating effects, particularly earlier in the day, will give you the best chance of developing a healthy relationship with foods of all kinds.

Weight-reducing strategies

- Start the day with a breakfast rich in protein, for example by including an egg, cheese, fish, meat or tofu, or nuts.
- Try to include some form of protein in every meal; the daily recommended value for women is 46 grams, and for men it is 56 grams. As an example, a can of tuna has around 30 grams of protein and a palm-sized chicken breast the same amount.
- Vegetables which are relatively high in protein include broccoli, spinach, cauliflower, asparagus, mushrooms and onions, so include them in your diet.
- If you are using protein to help aid weight loss, you can eat as many as 120 grams of it per day, so this may inform the kinds of foods you'll want to eat. (For further information on this topic go to www.thefatplanet.com).
- Add cottage cheese to your diet. It contains a surprisingly large amount of protein (28 g per cup!) while being very low in calories.
- Protein powders can also be blended into smoothies, which will help level out the blood-sugar-spiking effect of a sweet drink, and add satiating benefits.

8: Never shop for food when hungry

We demonstrated the importance of this advice ourselves, in an experiment involving sixteen young men and women. They participated in what they believed to be a study about teamwork. After being divided into two groups of eight they were kept busy, in different rooms, from early morning until early afternoon. One group was given only water while the second was supplied with snacks of fruits and nuts, each person's bag containing 370 calories. By lunchtime this meant that the first group was famished while the second, although hungry, felt much less urgency to eat. Each person was then given £10 and sent off to a nearby supermarket to buy whatever they wanted for lunch. Every item they selected was unobtrusively noted and the calorie and fat content recorded.

While we expected there to be a difference in food choice between the groups, we had not expected just how big a difference it would be. The total calories consumed at lunch by the group that had been provided with snacks amounted to an average of 700 per person. Individuals in the 'hungry' group, by contrast, ate an average of 2,500 each and also selected foods such as chocolate bars and crisps that were 400% higher in fat overall.[22]

It's clear from this that hunger can have a very significant effect on the kinds of foods we prefer at a given time.

Weight-reducing strategies
- Always eat before shopping for food. It doesn't have to be much, a healthy snack such as fruit or nuts containing around 200 calories, but less than 300, will be sufficient to take the edge off your appetite and significantly reduce your desire for HED foods.
- Be especially alert for special offers or 'bonus packs' of chocolates, crisps or sweets, which are often placed around the check-out area to tempt shoppers waiting to pay.
- Allow yourself a fixed amount of time to spend in the store. If you allow yourself the leisure of browsing you may end up convincing yourself you 'need' foods that you simply don't.

9: Be careful when eating in company

The tendency to consume more when eating in company is so predictable it is almost mathematical.[23] When eating with one other person we tend to eat around a third (35%) more than if alone. When dining with three others this increases to about 75% more, and as part of a group of seven or more up to 96% extra.[24]

Not only does eating with others cause overeating as a result of being distracted by the conversation, but diners tend to copy both the amount of food eaten by others and the speed of eating. In one study, confederates of the researchers acted as 'pacesetters' and ate at varying speeds with different groups of diners. The more food the pacesetters ate, the more the other diners also consumed. People also drink more alcohol in company than on their own. This not only increases the number of calories consumed (depending on the vintage, a glass of wine can contain up to 175 calories), but it also reduces awareness of how much has been eaten.[25]

Weight-reducing strategies

- Minimise the risk of overeating in company by having have a high-protein snack beforehand to reduce your hunger during the meal.

- Before starting the meal, ask for the bread basket and any high-calorie nibbles, such as olives, to be removed. If they are left, despite your best intentions, the chances are you'll end up eating them all.

- Decide how much to eat *prior* to the meal instead of during it.

- Order smaller quantities (e.g. half-size portions) and consciously avoid 'keeping pace' with others.

- If obliged, perhaps for business reasons, to eat frequently at restaurants, only consume half the entrée and forgo dessert.

- Remember to talk as much as possible! The more time you spend talking, the less time there will be for eating.

10. Beware of hidden persuaders

The ambience of the place in which food is eaten, its lighting, odour and noise, can also influence the amount people consume and cause diners to overeat. Researchers have found, for example, that while bright lighting decreases the length of time people spend over their meal, soft or warm lighting (including candle-light) generally causes people to linger longer and enjoy an unplanned dessert or an extra bottle of wine. This is why fast-food restaurants, where the desire is to move customers out of the premises as soon as possible, are typically very brightly lit, while exclusive eateries have dimmer lighting to encourage longer stays.[26]

Yet it might be surprising to learn that a relaxing environment actually inhibits eating overall. As an experiment, we converted two different floors of a large restaurant into an upmarket eatery and a fast-food diner respectively. The former had soft lighting, waitress service and played classical menu. The latter was brightly lit and played loud pop music. We found that although those dining in the relaxed atmosphere stayed for longer, they actually consumed fewer calories, an average of 749 per person, than those in the fast-food restaurant at around 949 each. This was most likely because by eating rapidly and mindlessly, the brain failed to respond in time to signals from the digestive system that sufficient food had been consumed. By encouraging a more leisurely approach to dining, the relaxed environment allowed feelings of fullness to be noticed earlier and more easily.

Weight-reducing strategies

- In the initial phases of implementing behavioural diet change, try to avoid restaurants. You'll be able to measure precisely how many calories you are consuming and won't have to contend with the difficult issues of portion size, ambience and temptation, which will likely make it harder to carry out your good intentions.

- Avoid any extras at restaurants; the pleasant setting and the mouth-watering descriptions on a menu might easily tempt you. This may go without saying, but stay away from desserts!
- Avoid alcohol; if you do drink wine, order it by the glass rather than surrendering to the temptation to linger over a bottle in a pleasant environment.
- In louder, brighter settings don't allow yourself to be rushed – enjoy the meal at your own pace.

As well as being more mindful of what and how you eat, there are four further steps one can take to encourage sustainable weight loss.

Weigh yourself weekly

There has been significant controversy about whether it is a good idea to monitor one's weight regularly. Arguments against doing so include the fact that if the reading is higher than expected it may lead to depression and so encourage comfort binge eating. Some people also worry that 'obsessing' about their weight makes it harder to achieve a healthy relationship with their body.

However, if you have a tendency to put on weight (and don't we all?), you must keep scales in your bathroom. No matter how painful it may be at first, weekly weight monitoring plays an important role in both losing weight and ensuring that it stays lost.

Researchers studied the behaviour of a cohort of successful dieters who had lost at least 30 kg, and maintained the loss for

2–5 years, from a sample obtained from the National Weight Control Registry. The five trends which emerged are not all that surprising. They are: maintaining at least 1 hour of physical activity per day, following a diet that is low calorie and low fat, consistency in eating habits between the weekends and weekdays and, finally, regular weekly weigh-ins.[27]

Improve your sleep

Lack of sleep has been associated with a higher incidence of obesity, hypertension, and higher total cholesterol and triglycerides levels. As we explained in Chapter 8, having insufficient sleep can lead to weight gain in a number of ways, not least by reducing self-control, leaving one vulnerable to impulsive eating.

In recent years the number of hours of undisturbed sleep the average person gets has fallen dramatically, in both the developed and developing world. To compound the problem, many people now work night shifts or, more damagingly still, split shifts, which disturbs their natural bodily rhythms.

During our waking hours our body temperature rises and falls; we warm up during the day, which makes us more alert, before cooling down in the evening, which causes us to feel sleepy. In most people, body temperature peaks around 6–7 p.m., and this is therefore when they naturally have the most energy. There is a small drop in body temperature around 3 p.m., which is when some people like to take an afternoon siesta, and then a large drop around 2–3 a.m. These are also times when snacking is most likely to occur.

Whilst we do not have the space to provide an extensive plan for improving one's sleep, here are some basic principles to follow. On our website (www.thefatplanet.com) you will find more detailed advice on improving your sleep.

Practical suggestions:

- Have a bedtime routine that you follow every night.
- Avoid any mentally stimulating activity within at least an hour of going to bed.
- Use your bed and bedroom primarily as a place to sleep; never try to work in bed.
- Many electronic devices such as mobile phones, laptops and tablets, give out a lot of blue light, which increases wakefulness. Switch digital products off at least two hours before settling down for sleep.
- If you fail to fall asleep within 15 minutes, get up and go into another room. Listen to music, read, or undertake any other relaxing activity for 5–10 minutes. As soon as you feel like you want to sleep, return to bed. If you aren't asleep within fifteen minutes, repeat the process. This prevents you becoming conditioned to stay awake when in bed, rather than falling asleep.

Take regular exercise

We are all frequently told of the benefits of exercise for our hearts and for our bodies in general; there is no need to repeat much of that information here. There is also a small amount of truth in the idea that if a person wants to lose weight the best thing they can do is 'move more'. However, the reason for this may not necessarily be the one you might expect.

It is certainly the case that exercise burns calories and, as we have seen, in the end it is excess calories that are the cause of weight gain. Having said this, the amount of exercise which most people feel able to take is unlikely to have a large direct impact on their weight because of this. Rather, it is the positive effect that exercise can have on mental health which can be particularly beneficial.

Exercise elicits the release of beta endorphins in the body –

hormones which activate opiate receptors and thereby produce feelings of wellbeing. Similar receptors are actually activated when we eat foods high in sugar and fat. Exercise can therefore prevent us from seeking the comforting high of unhealthy food because it provides a similar lift in mood itself.

We should also mention the results of a large population study published in the *European Journal of Preventive Cardiology*, which demonstrated that by taking regular exercise, maintaining a healthy diet, drinking alcohol only in moderation, and not smoking, people could reduce their risk of fatal and non-fatal cardiovascular disease by more than 57% and their risk of a fatal attack by 67%. When seven or more hours' sleep each night was added to these lifestyle factors, the overall protective benefit increased even further. Now there was a 65% lower risk of fatal and non-fatal cardiovascular disease and an 83% lower risk of fatal events.[28]

It may be something that you've heard a thousand times before, but if you find the time to take regular exercise you won't regret it – you'll feel better, both mentally and physically, which will go a long way to helping you maintain a healthy weight.

Be patient

Sustainable weight loss must be regarded as a long-term project rather than a quick fix. It takes up to two years for new and healthier eating habits to become established and for the appetite threshold to be reset. To expect to achieve this over any shorter period risks descending into what are, for many, all too familiar patterns of weight cycling and yo-yo dieting.

'In order to lose weight, you need to be motivated, and the best way to do that is through understanding what you're for, and how to change,' says writer James Fell. 'Not just your body, but your mind, your schedule and your reason for being. It's not just about the exercises you engage in and the diet you consume; it's about changing who you are. It's not a list of actions. It's someone you become. Don't just do this; be this.'[29]

In this part of our book we have focused on ways in which individuals can take ownership of and responsibility for their own body. Yet, external factors have no less important a role in encouraging overconsumption. In the final chapter, we examine the extent to which outside agencies such as NGOs, national governments, food companies and regulators need to become involved in combating the obesity pandemic.

CHAPTER 15

Planning for a Slimmer Planet

'We can't expect people to eat less if we continue serving them
too much food.'
Dr Deborah Cohen[1]

As we have seen, the obesity pandemic has arisen from multiple
causes, from the way food is grown, processed and sold, to the
ways in which consumption is advertised and marketed. It has
arisen from the relentless promotion of high energy-dense foods,
manufactured to be enormously rewarding, and as a result of the
techno-physiological revolution which has led to diets that, typically, are too high in calories and fats and too low in fruits and
vegetables.

'A century ago,' reports the US Department of Agriculture,
'nutritional problems centred on the inadequate intakes of certain
vitamins and minerals, resulting in nutritional deficiencies such
as rickets, scurvy, and beriberi. Now, nutritional problems in
America are driven by the discovery of strong links between nutrition and chronic diseases such as coronary heart disease, cancer,
and stroke.'[2] It has also been reported that Americans have *increased*
their annual food intake by over 190,000 calories per person since
the 1970s.[3] And like it or not, where America leads, other nations
will typically follow.

Finding our way back to a slimmer, healthier way of living
demands an understanding of consumer behaviour and the factors
influencing food choices.

What we eat is the end result of pulls in many directions, a response to the multiple forces creating the national nutrition environment. This is just as true today as it was two decades ago.

Instead of viewing obesity as the reflection of personal failure, we need to approach it as an issue of public health; the concern not just of the overweight but of society as a whole. The pandemic affects us all, both slim and obese, as a result of the social and financial burdens associated with its medical consequences.

We now recognise and accept the importance of food hygiene in safeguarding our health, and sanitary standards are enforced through a combination of education and regulation. Health inspectors have the legal powers to enter commercial premises to enforce stringent rules about the way food is stored, prepared and sold. Companies that fail to pass inspection can be closed down, their proprietors taken to court, fined and put out of business. At home and in school, children learn from an early age to wash their hands before eating, and to avoid food that looks or smells tainted.

No one, so far as we are aware, resents such sensible precautions or feels their personal freedom has been diminished by these steps taken to avoid the distressing and even potentially fatal sicknesses caused by bacteria such as salmonella or listeria.

We believe that the same combination of regulation and education must now be urgently adopted to combat the equally avoidable disease of obesity.* Given the mounting costs of the pandemic to individual and public health, this seems no more than common sense to us. Yet, after more than three decades of increasing obesity, the necessary stringent legislation has still not been put in place in Britain, in the US, or in other parts of the world.

There are several reasons for this.

One is fear of arousing hostility from press and public on the

---

* The diagnostic characterisation of obesity as a disease was made by the American Medical Association in June 2013.

part of democratically elected governments. They are wary of being accused of trampling on individual liberties by imposing a 'nanny state' and patronising responsible adults who are capable of exercising their own judgement over what and how much they choose to eat. When a simple move by Mayor Bloomburg to limit soda (fizzy drink) size to 16 oz (a pint) was put forward, also known as the 'Portion Cap Ruling', it was to the relief of millions working in public health and also citizens throughout the US. At long last, it looked as though there was some leadership supporting the transition towards a healthier eating environment. Ultimately, we do need to start building concrete barriers to limit our boundless access to food and expanding appetite. Unfortunately, after months of regulatory battles, the New York State of Appeal issued the final decision, ruling: 'We hold that the New York City Board of Health, in adopting the "Sugary Drinks Portion Cap Rule," exceeded the scope of its regulatory authority. By choosing among competing policy goals, without any legislative delegation or guidance, the Board engaged in law-making and thus infringed upon the legislative jurisdiction of the City Council of New York.'[4]

The question remains: if all parties and citizens are united in their concern for their health, and we can't start putting limits on portion size, what chance do we stand of reforming our food environment to allow for greater health?

The enormous financial clout and high-level lobbying power of Big Food is in itself a serious obstacle to effective new regulation. The industry's capacity for watering down or blocking any legislation likely to harm their profitability is as awesome as it is shocking.

However, more optimistically we should remember that in many countries, the same battles have been fought and largely won against the equally powerful tobacco and alcohol lobbies. Just as governments have realised the need to bring these two pleasurable, but unhealthy products under regulatory control, so too must they now adopt a similar approach to the foods which fuel obesity. Indeed, in this book we hope we have convincingly made the case that HED foods can prove just as dangerous as any hedonically excessive behaviour, such as smoking or drinking – a view most leading obesity researchers and neuroscientists share.

In addition to new legislation, bringing the obesity pandemic under control will require educational campaigns sufficiently imaginative and well-funded to match the billions spent on advertising and marketing by Big Food. Starting in nursery school, we need to bring about fundamental changes to the way people relate to food and to the obesogenic environment in which they live. It's a vast and daunting task. 'We haven't scratched the surface of what we need to do,' Deborah Cohen of the RAND Corporation told us. 'Some of the solutions seem so obvious and so simple, but nobody is talking about them.'[5]

She is right. And while there can be no single solution to a problem that is, as we have seen, extremely complex and multifaceted, we have identified a number of measures that would help our planet become slimmer and healthier.

Before considering these, however, let's review two currently favoured approaches – the search for a magic bullet slimming pill, and the formation of public–private partnerships.

Overweight? Pop a Pill

There's no business like the obesity business. Dieting, surgery, gyms, spas, pills, potions and lotions, personal trainers and dieticians are just some of the means people employ in what has become a multi-billion pound battle against the bulge.

No sooner has some high-profile TV medic (or even a celebrity with little if any nutritional knowledge) endorsed the latest fad diet than millions rush to follow it. Out of desperation to lose weight, people will also follow any 'guaranteed' quick and easy diet or slimming book, DVD or television show. They will believe almost any research and swallow any nostrum if marketed persuasively, especially when the claims being made are backed up with apparent medical opinion.

'You may think magic is make-believe, but this little bean has scientists saying they found a magic weight-loss cure for every body type,' announced American cardiologist Dr Mehmet Oz, when describing the supposed fat-shedding benefits of green coffee bean extract.[6] He made this statement after Indian-based researchers, funded by the US company Applied Food Sciences, reported that participants who had been fed the extract had lost around 18 lb and 16% of fat in just one week. Eager not to miss out on such an effortless way to lose weight, and encouraged by the endorsement of respected medics like Dr Oz, people rushed to buy this latest answer to their slimming prayers – it sounded almost too good to be true.

And it was.

The results had been fabricated. A subsequent investigation found those taking the green coffee extract had, over the course of a week, lost an average of only 2 lb while participants given a placebo had lost 1 lb. While the beans seemingly had some effect, it was nowhere near as dramatic or as significant as the marketing and advertising had implied.

After months of deliberation, Applied Food Sciences reached a $3.5 million settlement with the Federal Trade Commission.[7] Summoned before a Senate Commerce Committee, Dr Oz was told by Senator Claire McCaskill: 'I don't get why you need to say this stuff, because you know it's not true. So why, when you have this amazing megaphone and this amazing ability to communicate, would you cheapen your show?'[8] Sadly this isn't a unique case. The obesity business peddles more extravagant promises, contains more hucksters and creates more dashed hopes than any other we know. It's a fiercely competitive market in which, as the Victorian poet Matthew Arnold wrote: 'Ignorant armies clash by night.'

Still, many pharmaceutical companies have invested and continue to invest millions in the search for the perfect slimming drug. Often with nothing to show for their investment, as in the case of leptin (see Chapter 5). There are, however, some drugs that can bring about modest weight loss – an average of around 5%. Available on prescription only, they include topiramate and contrave, both of which can help suppress binge eating.[9]

However, because the interactions between the central nervous system and the endocrine system are so complex, it could be decades before any drugs that have a really significant effect on obesity come on to the market. Even were such a drug to be developed, should society really favour 'medicalising' a condition caused, at least initially, by living in an obesogenic environment? While medical interventions, including surgery, may well prove necessary when lives are threatened by excessive obesity, they surely ought not to become the universal panacea for everyone who needs to shed a few pounds.

Once such a pill was available there would be such a demand for this 'no effort' way to lose weight that doctors would be overwhelmed, not only by patients with serious, weight-induced health conditions, but also individuals anxious to drop a few pounds for purely cosmetic reasons. Furthermore, such a drug would also enable people to continue to eat unhealthily, so even if they were thin, they might still be denying their body essential nutrition. And finally, although external factors are a strong cause, we arguably shouldn't be completely absolved of responsibility for our own welfare. Adopting a pharmacological 'cure' for something as intimately tied to survival as eating might well lead to greater problems in the long run.

The fact that new drugs are frequently misused and overprescribed is something that physician Allen Frances addresses in his book *Saving Normal*.[10] 'Primary care physicians,' he notes, 'are prescribing potentially dangerous medications, outside their competence, for people who should not be taking them. Proof again that drugs that are too easy to give and too easy to take will be taken far too often, especially when lots of money is behind them.'[11] Frances is talking about the drugs used to treat bipolar affective disorder, but the same warning applies to the possibility of treating weight problems through medication.

Far better, in our view, to tackle the environmental and psychological causes of obesity than seek a quick fix via the prescription pad.

Public–Private Partnerships

Recognising that the food industry must be part of the solution to the obesity pandemic, a growing number of Big Food companies are joining forces with government and related agencies and making pledges to improve the status of our obesogenic environment.[12] The rationale is simple; with public awareness of obesity rising, it would be commercially prudent for these companies to be seen to be offering lower-calorie, nutrient-dense foods. Big Food also knows that making a visible contribution to tackling obesity could help fend off the costs and restrictions of formal legislation.[13]

As a result, there are all kinds of public–private partnerships to tackle obesity in the world today. In the US, Michelle Obama's 'Let's Move!' campaign to end childhood obesity is defined as a 'public–private partnership that, for the first time, sets national goals to end childhood obesity in a generation.'[14] It is perhaps the most prominent campaign of its kind in all of history.

At a global level the World Health Organization's 2008–2013 Action Plan for Non-Communicable Diseases made specific requests to the private sector to become a partner in their world-wide weight reduction efforts.[15]

The French-based partnership, EPODE International Network (Ensemble Prévenons l'Obésité Des Enfants) is a not-for-profit organisation operating in 15 countries and involving 150 million people. Supported by the civil and corporate sector, its aim is to reduce childhood obesity through community-based programmes.[16]

On a smaller scale, the Diet and Health Research Industry Club (DRINC) is a UK-based organisation, funded both publicly and privately, which seeks to innovate new, healthier food and drink products.[17] Meanwhile another UK-based partnership, Responsibility Deal, involves around 150 companies pledged to improve the current food environment.[18]

Undeniably these partnerships do a great deal of good work overall. However, it has been argued that some companies are involved in them as a way of diverting responsibility for obesity from obesogenic big business and ensuring that it remains thought of as an issue of individual responsibility. Based on what little is known about their track record to date, such charges can be levelled against a number of these initiatives.[19]

For one thing, because 'partnership' is only loosely defined, the standards of what constitute models of best practice may be compromised.[20] While many private-sector players have announced plans to improve food composition, nutrition labelling, and reduce marketing to children, specialists dispute the extent to which they have delivered on these promises.[21] Independent studies have reported that the current self-regulatory framework – which tends to put pressure on the individual rather than looking at the soci-etal factors that contribute to overeating – has not led to improvements in either diet or health outcomes.[22]

The Access to Nutrition Index (ATNI), funded through partner-ships between the Global Alliance for Improved Nutrition, Bill and Melinda Gates Foundation and the Wellcome Trust, evaluates food and beverage manufacturers on their commitment to nutrition, disclosure, and efforts targeted towards obesity and food scarcity.[23] Using a combination of public data, market research and direct interviews with company spokespeople, the ATNI have developed a scoring system to rank corporations according to the extent they have lived up to their promises. The ATNI is anything but a public–private partnership. It relies on scientific grant bodies to fund investigations, thereby eliminating any potential conflict of interest.

Their initial report, published in 2013, found that most corporate efforts were feeble at best, failing to measure up to the commitments that had been made. Moreover, it found there was a lack of transparency in the ways the initiatives improved health outcomes.[24]

So while they undoubtedly have some value, overall public–private partnerships are arguably a distraction, serving to reassure people that appropriate steps are being taken to tackle the obesity crisis and thereby prevent the more radical actions that are actually needed.

We believe there are at least seven strategies that governments around the world must adopt as a matter of urgency. None will be popular. All will face opposition from well-funded and powerful commercial interests, but each would offer enormous health benefits, and we must begin as soon as possible to improve the prospects of future generations.

Seven Steps to Reduce Obesity

One: improve food education

There is a need to move beyond formulaic instructions such as 'eat less, move more', and beyond nagging people to eat more fruits and vegetables. While these are key messages, it is equally important to explain the adverse health consequences of over-eating. To have any chance of being successful, these messages need to be communicated with the same persuasive power and reach as Big Food, and that will mean significant spending.

It's important to begin imparting ideas about healthy eating to children at as early an age as possible. They need to be taught about the metabolic syndrome and the risks it poses not just to their physical and mental health, but to their self-image and sense of self-worth. The messages should be delivered in ways as compelling as that employed to promote HED snacks and drinks.

This can be done. In 2006, for example, the British Heart Foundation ran a campaign depicting a young girl drinking from a bottle of cooking oil with the slogan: 'What goes into crisps goes into you.'[25] This campaign was based on the discovery that British children, with their pack-of-crisps-a-day habit, were consuming upwards of 9 litres of oil a year. Attention-grabbing advertising has also proved successful in reducing smoking among adolescents. In the late 1990s, the 'Truth' campaign was established in Florida to help educate teens about tobacco use. Created by the advertising agency Crispin Porter & Bogusky, 'Truth' advertisements were featured on MTV during prime time, and funded to the tune of $15 million by the government of Florida, using money raised by legal action against the tobacco companies.[26] One of the best Truth campaign advertisements featured a withered old executive wearing a bikini and smoking, with the tag line, 'No wonder tobacco executives hide behind sexy models.' To the left, in a warning box typical of those featured on cigarette packets, are the words 'WARNING: Their brand is lies. Our brand is truth.' It's a fun and quirky visual, and diminishes the sexy edge that cigarette advertisements have traditionally relied upon.

A similar approach needs to be taken with advertising aimed at combating obesity and promoting healthy eating – the messages conveyed have to be as fun and contemporary as those that are being used to sell unhealthy foodstuffs. Having said this, care needs to be taken not to overemphasise an idealised appearance as a primary reason for eating healthily. Campaigns taking that approach run a serious risk of encouraging disordered eating and body dysmorphia. Children should not be made to feel inadequate through being asked to conform to ridiculous ideals of physical attractiveness; in its own way this is every bit as destructive as anti-obesity messages that suggest being overweight is a choice and is largely caused by personal greed. Instead young people need to be helped to understand the types of food their body needs – and does not need – in order to function healthily and effectively.

Two: restrict HED food advertising to children

There should be a ban on showing commercials for foods and snacks with a high fat, sugar or salt content on children's television. Such a ban could be voluntarily carried out by the channels, but realistically would probably have to be legally enforced. This measure would encourage manufacturers to focus more on the nutritional value of their breakfast products in order to allow them to continue to be advertised during such times. Similarly, promotion of HED foods aimed at children by celebrities should be prohibited. As we saw in Chapter 10, such adverts and promotions have been proven to prime children to prefer particular foods. In order to give them the best possible chance of developing healthy eating habits we need to prevent them, as far as possible, from being indoctrinated in this way.

Three: adopt a traffic-light system for food labelling

Foods should be clearly categorised according to their nutritional values. One quick and easy way of doing this would be to adopt a traffic-light system – in fact, one was mooted for UK shops but never finally introduced. Here, red would indicate food with little if any nutritional value, amber (or yellow) would be for food with some value, and green would indicate food with high nutritional value.

Such a system would enable busy customers to quickly make informed choices about the food they are buying. It could be especially helpful to parents needing to determine which foods were best and worst for their children when out shopping with them (and probably being asked to purchase particular brands). To expect the parents of small children to find the time to read the small print on a food label describing its contents is simply not realistic.

Knowing their foods would be categorised in this way would also encourage food manufacturers to reformulate HED products. In much the same way that Hollywood producers anxious to make their films available to a juvenile audience will edit them in such a way as to ensure a child-accessible certification, it seems probable that they would do all they could to pull products into the amber and green categories to make parents feel comfortable buying them for their children.

It would also provide a simple way of determining the times at which such products could be advertised on television. Those with a red designation might be limited to slots after nine p.m., for example.

At the time of writing, Coca-Cola has announced plans to adopt a traffic-light system for beverages sold in the UK. In accord with the Responsibility Deal (an act made to encourage businesses and other influential organisations to enhance health behaviours), Coca-Cola will adopt red/yellow/green packaging labels to aid the customer in making decisions to reduce caloric intake. This change is perhaps indicative of an important inflection point; that is, the start of a change in attitudes on the part of some companies. A paradigm shift is certainly needed if the goal of transparent food policy is to be achieved. Critically, pressure is needed if change is to occur; our health merits tighter controls on what companies are allowed to produce and how those products are marketed. Coca-Cola's decision to embrace the traffic-light system was well thought out, and is a welcome development. There is, however, still a very long way to go.

Four: ban high energy-dense foods from schools

Keeping in mind the fact that there are striking similarities between brain responses to hyper-palatable foods, and brain responses to addictive drugs, we need to start treating highly processed food as the potentially problematic substance it really is.[27] 'Sugar may not pose the clear addictive characteristics of illicit drugs such as cocaine and heroin', comments a Credit Suisse research report on the sugar industry. 'But to us it does meet the criteria for being a potentially addictive substance.'[28]

In the light of these findings, school shops must stop selling nutritionally poor, hyper-palatable snacks or sugar-based drinks, and machines vending these products must be banished from school property. Children should be forbidden to leave the premises during the school day to top up on fast food, sweets and crisps from the nearest shop.

Extreme as it may sound, the law should arguably be changed to prohibit shops from selling HED foods to anyone under a certain age, in the same way that it is illegal in many countries for retailers to sell alcohol or tobacco products to underage children. Once introduced, these statutes should be policed and enforced just as rigorously.

Without doubt the Big Food lobby would fight hard to prevent such measures and maintain the status quo. But, as Deborah Cohen – who served on several advisory panels for the National Institute of Health in the US – told us: 'Nobody likes this idea. When they first hear it, they find it odd. We have a misconception that individuals are always in full control of everything they do, including what and how much they eat. Until people understand they're vulnerable to marketing and to convenient, easily accessible food, there won't be support for protecting people against an obesogenic environment. Once people recognize they're vulnerable, they'll be really happy to have some help. The private sector has learned how to manipulate people, and it's the government's job to protect people from manipulation and the risks of chronic diseases.'[29]

Five: reduce the opportunities for impulse purchases

No government should, or could, try to stop people from eating a bar of chocolate a day if that's what they choose to do. However, governments surely have an obligation to protect their populations against marketing practices which, as Deborah Cohen puts it, 'force you to confront chocolate bars every day.'[30]

This is an important distinction. As we explained in Part Four, the spend on food marketing is second only to the tobacco industry, with $1.79 billion going to advertise junk food aimed at children, compared to $280 million for healthier foods.[31] We now know that those most vulnerable to the rewarding effects of food are also more sensitive and reactive to food-related imagery.[32]

In the UK, restrictions on impulse marketing have been addressed better than elsewhere in the industrialised world. Companies cannot direct television advertising for high fat, salt and sugar (HFSS) foods to children. The Food Standards Agency identifies which foods constitute as HFSS. Even foods that are not categorised as HFSS must abide by a set of rules established for television advertising.

A small, but important step has also been the ban on 'treats at the till' such as chocolate bars.[33] This approach to limiting temptations to British consumers was thwarted by smaller retail stores, but embraced by big stores such as Tesco and Aldi.[34]

These kinds of measures – and additional ones – should become standard in all countries.

Six: encourage companies to produce healthier food

Hank Cardello, head of the Obesity Solutions Initiative at the Washington DC-based think tank, the Hudson Institute, has the task of demonstrating to companies the measurable financial benefits of offering healthier food to the public. Perhaps surprisingly, given all that we have said, he reports that it is not a task he is finding especially difficult.

'Most growth in the food industry today is coming from healthier, lower calorie options,' he told us. 'Simply put, it's just good business.'[35] Healthier eating is, indeed, growing in popularity amongst certain groups in society.[36] As reported in Chapter 12, farmers' markets and those offering whole foods and fresh produce are becoming an important development within the food industry. However, this is not a general trend across the whole of the populace. It's true that certain segments of the market can and will pay a premium for healthier, lower calorie, palatable foods; the challenge remains in making this food available to a wider variety of people.

Logistics present a significant problem in the distribution, and particularly creation of fresh produce. As we detailed in Chapter 12, not only are fresh fruits and vegetables vulnerable to spoilage, certain climates are inappropriate for growing them. Attempting to grow crops in climates without the appropriate soil conditions leads to increased waste and use of fertilisers.

This could, however, all be changed, with food grown hydroponically under cooler LED illumination. This could lead to a significant reduction in the energy needed, up to 85% according to some reports; one tenth of the amount of water used by conventional growing operations and carbon dioxide falling by some two tons a month.[37]

'By growing our crops vertically, we are able to pack more plants per acre than we would have in a field farm, which results in more harvests per year,' says Robert Colangelo, founding farmer/president of Green Sense Farms. 'We produce little waste, no agricultural runoff and minimal greenhouse gasses because the food is grown where it is consumed.'[38]

This kind of technology makes economical production of healthy foods a real possibility – it is to be hoped that companies do not squander that opportunity by opting instead to try to squeeze profit margins still further.

Seven: encourage the manufacture of *low-calorie* hyper-palatable foods

There have been attempts in some countries to lower people's caloric intake. For example, the Healthy Weight Commitment Foundation has compelled US businesses to reduce calories across all products offered to the American public. To an extent, this has been successful, with 6.4 trillion calories being removed from American food over the last three years.[39] It is also an example of the food industry proactively trying to create palatable, healthier alternatives to less nutritious foods. Sodexo, the French conglomerate, for example, collaborated with ConAgra to create 'Ultragrain', a product that tastes like white flour but is in fact made from whole grains.[40]

While this is a step in the right direction, the food industry is under pressure to continue to perform, so we as consumers have a responsibility to choose healthier foods to demonstrate that there is demand.

The strategies described above would not be especially irksome for either the companies who produce, market and retail food, nor for the customers who buy them, were we to take a different attitude towards the way we eat.[41]

As with other changes introduced to improve health and reduce the risk of premature death, initial resistance can be expected. It happened when it came to improvements in food hygiene, the introduction of building regulations and the ban on smoking in public places. Although fiercely objected to at first, these were soon generally accepted.

And what is the alternative? Over the next twelve months, the obesity pandemic will take the lives of almost three million people and undermine the health of billions more.[42] Obesity is estimated to present as much of a financial burden as either smoking or armed violence and terrorism, with costs estimated at around $2 trillion globally.[43] In light of this enormous cost to health-care systems, both industry and people must work together to minimise this cost. Families should not have to bear the pain of such preventable losses. Individuals should not have to suffer years of pain and discomfort resulting from mindless overeating, nor society the soaring public health costs of coping with this crisis.

We must abandon our deeply ingrained prejudices about obesity and come to understand that we are all physiologically and psychologically ill equipped to resist the temptations of our obesogenic environment. We must return to viewing food as an essential fuel rather than a recreation. We must curb the power of big businesses to market high energy-dense foods and encourage them to produce nutritious alternatives, supported by technological advances that make this course economically viable. And yes, we must all take personal responsibility for our own health.

If we do not, we risk devastating consequences for everyone on this increasingly fat planet.

Chapter References and Notes

Introduction

1 http://familyfitness.about.com/od/motivation/a/michelle_obama_quotes.htm

2 Sassi, F., et al., 2009. 'The Obesity Epidemic: analysis of past and projected future trends in selected OECD countries'. *OECD Health Working Papers*, 45; Lobstein, T., Baur, L. & Uauy, R., 2004. 'Obesity in children and young people: a crisis in public health'. *Obesity Reviews*, 5 (suppl 1), 4–104.

3 Blythman, J., 2006. *Bad Food Britain: How a Nation Ruined its Appetite.* London: Fourth Estate, p. 247; WHO. *Reducing Risks, Promoting Healthy Life.* Geneva: World Health Organization, 2005.

4 Ng, M., et al., 2014. 'Global, regional, and national prevalence of overweight and obesity in children and adults during 1980–2013: a systematic analysis for the Global Burden of Disease Study 2013'. *The Lancet*, 384 (9945), 766–81; Finucane, M. M., et al., 2011. 'National, regional, and global trends in body-mass index since 1980: systematic analysis of health examination surveys and epidemiological studies with 960 country-years and 9.1 million participants'. *The Lancet*, 377 (9765), 557–67.

5 Diabetes in the UK, 2012. 'Diabetes in the UK, Key Statistics on Diabetes'. In Diabetes.org.uk, ed. http://www.diabetes.org.uk/Documents/Reports/Diabetes-in-the-UK-2012.pdf.

6 Boyle, J.P., et al., 2001. 'Projection of Diabetes Burden Through 2050: Impact of changing demography and disease prevalence in the U.S'. *Diabetes Care*, 24 (11) 1936–40.

7 Stephenson, T., 2014. *Measuring Up: The Medical Profession's Prescription for the Nation's Obesity Crisis.* London: Academy of Royal Colleges.

8 Webber, L., 2014. 'The European Obese Model: the shape of things to come'. In: Forum UH, ed. 2014.

9 Flegal, K. M., et al., 2010. 'Prevalence and trends in obesity among US adults, 1999–2008'. *Journal of the American Medical Association*, 303 (3), 235–41.

10 Webber, L., et al., 2014. 'The future burden of obesity-related diseases in the 53 WHO European-Region countries and the impact of effective interventions: a modelling study'. *BMJ Open* 4, ee004787

11 Ibid.

12 Popkin, B. M., 2001. 'The nutrition transition and obesity in the developing world'. *Journal of Nutrition*, 131 (3), 871S–873S; Grover, A., 2014. 'Unhealthy foods, non-communicable diseases and the right to health'. In UNG Assembly, ed. *Report of the Special Rapporteur on the right of everyone to the enjoyment of the highest attainable standard of physical and mental health*, 1–68.

13 2014. 'Chubby little emperors', *The Economist*, 14 June, p. 68.

14 Popkin, 'The nutrition transition'.

15 Wang, Y. & Lim, H., 2012. 'The global childhood obesity epidemic and the association between socio-economic status and childhood obesity'. *International Reviews on Psychiatry*, 24 (3), 176–88.

16 Ogden, C. L., et al., 2012. 'Prevalence of obesity and trends in body mass index among US children and adolescents, 1999–2010'. *Journal of the American Medical Association*, 307 (5), 483–90.

17 De Onis, M., Blossner, M., & Borghi, E., 2010. 'Global prevalence and trends of overweight and obesity among preschool children'. *American Journal of Clinical Nutrition*, 92 (5), 1257–64.

18 Overweight and Obesity viz, application sponsored by the Bill & Melinda Gates Foundation. http://vizhub.healthdata.org/obesity/ accessed September 2014.

19 Diabetes.org.uk. 2013. 'Number of people diagnosed with diabetes reaches three million'. http://www.diabetes.org.uk/About_us/News_Landing_Page/Number-of-people-diagnosed-with-diabetes-reaches-three-million/ accessed March 2014.

20 Wang, Y. C., et al., 2011. 'Health and economic burden of the projected obesity trends in the USA and the UK'. *The Lancet*, 378 (9793), 815–25.

21 Ibid.

22 Lu, B., et al., 2014. 'Being overweight or obese and risk of developing rheumatoid arthritis among women: a prospective cohort study'. *Annals of Rheumatic Diseases*. Epub ahead of Print.

23 De la Monte, S. M. & Wands, J. R., 2008. 'Alzheimer's disease is type 3 diabetes – evidence reviewed'. *Journal of Diabetes Science and Technology*, 2 (6), 1101–13.

24 Ibid; De la Monte, S. M. & Tong, M., 2013. 'Brain metabolic dysfunction at the core of Alzheimer's disease'. *Biochemical Pharmacology*, 88 (4), 548–59.

Part One – The Obesity Blame Game

1 A food's caloric value indicates the amount of heat it produces. Technically a calorie is the unit of heat required to raise the temperature of 1 g of water from 15

to 16 degrees Celsius (centigrade). Because this is a very small amount of heat, the convention is to refer to the kilocalorie (kcal) equal to 1,000 calories or the large calorie, usually with a capital C.

2 Tom Colicchio (American celebrity chef), interview CNN Eatocracy http://eatocracy.cnn.com/category/news/celebrity-chefs/tom-colicchio/

Chapter 1 – What It Feels Like to Be Fat

1 Diane Carbonell, 2013. 'I Could Not Escape Mirrors: Seeing Myself Without Seeing', Fit to the Finish blog, 8 November. http://blog.fittothefinish.com/2013/11/i-could-not-escape-mirrors-seeing-myself-without-seeing

2 For many people, especially those struggling with their weight, the terms 'obese' and 'obesity' can appear insensitive and abusive. Indeed, the words themselves – from the Latin *obesus* 'to eat' – have been criticised by some health experts as 'derogatory'. Among these, somewhat paradoxically, were the authors of a report published in 2012 by the UK's National Institute for Health and Care Excellence, which cautioned that: 'The term "obesity" may be unhelpful among some communities – while some people may like to "hear it like it is", others may consider it derogatory.' 'Obesity: Working with Local Communities', NICE public health guidance 42 (November 2012), www. guidance.nice.org.uk/ph42. Not everyone agrees. Mr Tam Fry, spokesman for the National Obesity Forum, commented: 'This is extremely patronising. They should be talking to people in an adult fashion. There should be no problem with using the proper terminology. If you beat around the bush then you muddy the water.' [Adams, S., 2012. 'Obesity a "derogatory" word, says NICE', *Daily Telegraph*, 8 May. http://www.telegraph. co.uk/health/healthnews/9252311/Obesity-a-derogatory-word-says-Nice.html] Obesity is a well-defined physical state established as a standard by the World Health Organization.

3 Carbonell, 2013, 'I Could Not Escape'.

4 Hawksworth, E., 2013. 'What It's Really Like to Live as a Fat Person Every Day'. *Huffington Post*, 18 October. http://www.huffingtonpost.ca/elizabeth-hawksworth/being-overweight_b_4116510.html

5 Cited in: Cloake, F., 2013. 'Our Big Fat Fear', *New Statesman*, 17–23 May, pp. 25–28.

6 From Rorer, Sarah Tyson Heston, 1898. *Good Cooking – Ladies' Home Journal Household Library*, cited in Foxcroft, L., 2011. *Calories and Corsets: A history of dieting over 2,000 years*. London: Profile Books Ltd. Sarah Tyson Rorer (1849–1937) was America's first true dietician. In 1878, after attending a few medical lectures and spending six months at cookery classes, she opened the Philadelphia Cooking School. Her students were taught about proteins and carbohydrates, but nothing about either vitamins or calories. Over the school's thirty-three years, some 400 students graduated as fully qualified diet specialists.

7 'Adult obesity and socioeconomic status', National Obesity Observatory Data Factsheet, September 2012.

Chapter 2 – 'Hey fatty, get off this train!'

1 http://www.stroustrup.com. Professor Bjarne Stroustrup was actually referring not to the obesity pandemic but computer software, but his perceptive comment clearly has application to many of the problems confronting us today.

2 Kevan, P., 2009. 'Beaten up by train thug for being fat.' *Metro*, October 19.

3 Winterman, D., 2009. 'Why Are Fat People Abused?' *BBC News Magazine*, 29 October. http://news.bbc.co.uk/1/hi/8327753.stm, downloaded August 2014.

4 Ibid.

5 Americans with Disabilities Act of 1990 (July 1990), 104. Stat. 327. Cited from Gilman, S.L., 2004. *Fat Boys*. Lincoln: University of Nebraska Press, 14–15.

6 Winterman, 'Why Are Fat People Abused?'.

7 Ibid.

8 Puhl, R. M. & Heuer, C. A., 2010. 'Obesity Stigma: important considerations for public health', *American Journal of Public Health*, 100 (6), 1019–1028.

9 Rees, R. W., et al., 2014. 'It's on your conscience all the time': a systematic review of qualitative studies examining views on obesity among young people aged 12–18 years in the UK. BMJ open. 4 (4), e004404.

10 Ibid.

11 Ibid.

12 Ibid.

13 Degher, D. & Hughes, G., 1999. 'The adoption and management of a "fat" identity.' In D. Maurer and J. Sobal, eds. *Interpreting Weight: the Social Management of Fatness and Thinness*. New York: Aldine de Gruyter, 11–27.

14 Campbell, Ian, cited in Winterman, 'Why Are Fat People Abused?'

15 Rand, C. S. & Macgregor, S. M., 1990. 'Successful weight loss following obesity surgery and the perceived liability of morbid obesity'. *International Journal of Obesity*, 15 (9), 577–9.

16 Mendez, J. & Keys, A., 1960. 'Density and composition of mammalian muscle'. *Metabolism* 9, 184–8; Ross, R., et al., 1985. 'Adipose tissue volume measured by

magnetic resonance imaging and computerized tomography in rats'. *Journal of Applied Physiology* 70 (5), 2164–72.

17 Eknoyan, G., 2008. 'Adolphe Quetelet (1796–1874) – the average man and indices of obesity'. *Nephrology Dialysis Transplantion* 23 (1), 47–51; Quetelet, A., 1832. *Nouveaux Memoire de l'Academie Royale des Sciences et Belles-Lettres de Bruxelles.* t. VII; Quetelet, A., 1842. *Treatise on Man and the Development of his Faculties.* Reprinted in 1968 by Burt Franklin, New York. In this book he developed the concept of *l'homme moyen* ('average man'), whom he characterised by the mean values of measured variables that follow a normal distribution.

18 Keys, A., et al., 1972. 'Indices of relative weight and obesity'. *Journal of Chronic Disease,* 25 (6–7), 329–43.

19 Peltz, G., et al., 2010. 'The role of fat mass index in determining obesity'. *American Journal of Human Biology,* 22 (5), 639–47.

20 Shiwaku, K., et al., 2004. 'Overweight Japanese with body mass indexes of 23.0–24.9 have higher risks for obesity-associated disorders: a comparison of Japanese and Mongolians'. *International Journal of Obesity,* 28 (1), 152–8.

21 Bei-Fan, Z., Cooperative Meta-Analysis Group of the Working Group on Obesity in China, 2002. 'Predictive values of body mass index and waist circumference for risk factors of certain related diseases in Chinese adults: study on optimal cut-off points of body mass index and waist circumference in Chinese adults'. *Asia Pacific Journal of Clinical Nutrition,* 11 (Supplement 8), S685–S693.

22 Devlin, K., 2009. 'Top 10 Reasons Why the BMI is Bogus', National Public Radio (July 2004). http://www.npr.org/templates/story/story.php?storyId=106268439 accessed November 20 2014.

23 Centers for Disease Control and Prevention, 'Healthy Weight – it's not a diet, it's a lifestyle!' http://www.cdc.gov/healthyweight/assessing/Index.html accessed September 2013.

Chapter 3 – Fashions in Fat

1 Wallis Simpson, quote retrieved from http://www.answers.com/Q/Who_originally_said_You_can_never_be_too_rich_or_too_thin

2 Clarke, D. T. D., 1981. *Daniel Lambert,* Leicestershire Museum Publications, 23 (3).

3 Bondeson, J., 2004. *The Two-Headed Boy and Other Medical Marvels.* Ithaca: Cornell University Press.

4 Stearns, P. cited in, Smith, D., 2004. 'Demonizing Fat in the War on Weight', *New York Times,* 1 May, http://www.nytimes.com/2004/05/01/arts/demonizing-fat-in-the-war-on-weight.html accessed January 14, 2015.

5 Tafrate, P., 2008. 'The New England Fat Men's Club'. *History*, July/August, 47–9.

6 Gilman, S. L., 2004. *Fat Boys: A Slim Book*. Nebraska: University of Nebraska Press.

7 Payne-Palacio, J. & Canter, D. D., 2000. *The Profession of Dietetics*. Baltimore: Lippincott Williams & Wilkins, 5.

8 Howard, C., 2012. 'The Big Picture', *The Economist*, 15 December, http://www.economist.com/news/special-report/21568065-world-getting-wider-says-charlotte-howard-what-can-be-done-about-it-big accessed January 15, 2015.

9 Gilman, *Fat Boys*.

10 Jay, M., et al., 2009. 'Physicians' attitudes about obesity and their associations with competency and specialty: A cross-sectional study'. *British Medical Care Health Services Research* 9 (106).

11 Foster, G. D., et al., 2003. 'Primary care physicians' attitudes about obesity and its treatment'. *Obesity Research* 11 (10), 1168–77.

12 Ibid.

13 Block, J. P., DeSalvo, K. B. & Fisher, W. P., 2003. 'Are physicians equipped to address the obesity epidemic? Knowledge and attitudes of internal medicine residents'. *Preventative Medicine* 36 (6), 669–75.

14 Jay et al., 'Physicians' attitudes'.

15 Elkin, S., 2013. '5:2 is just the latest: Britain's diet industry is worth £2 billion, so why do we buy into it?' *Independent*, 1 August. http://www.independent.co.uk/voices/comment/52-is-just-the-latest-britains-diet-industry-is-worth-2-billion-so-why-do-we-buy-into-it-8737918.html accessed 3 September 2014.

16 Reisner, R., 2008. 'The Diet Industry: A Big Fat Lie'. *Bloomberg Business Week*. http://www.businessweek.com/debateroom/archives/2008/03/the_diet_industry_a_big_fat_lie.html accessed September 3, 2014.

17 *The Guardian*, 2009. Datablog downloaded from http://www.theguardian.com/news/datablog/2011/jul/22/plastic-surgery-medicine#data accessed 20 April 2014, 3 September 2014.

18 Campos, P. cited in, Smith, 'Demonizing Fat'.

19 Campos, P., 2004. *The Obesity Myth: Why America's obsession with weight is hazardous to your health*. New York: Gotham, 233.

20 Wadden, T., & Sarwer, D., 1996. 'Behavioral treatment of obesity: new approaches to an old disorder'. In: Goldstein, D., ed. *The Management of Eating Disorders*. Totowa, NJ: Humana Press.

21 Ng, M., et al., 2014. 'Global, regional, and national prevalence of overweight and obesity in children and adults during 1980–2013: a systematic analysis for the Global Burden of Disease Study 2013'. *The Lancet*, 384 (9945), 766–81.

22 Stearns, in Smith, 'Demonizing Fat'.

23 Germaine, G., 2007. 'Well done, Beth Ditto. Now let it all hang out'. *The Guardian*. 31 May, http://www.theguardian.com/music/2007/may/31/news. germainegreer, accessed January 15 2015; CNN, 2006., 'The insider's guide to Beth Ditto'. *The Briefing Room*, 24 November, available from http://edition.cnn.com/2006/ SHOWBIZ/Music/11/24/tbr.ditto/index.html, retrieved June 2014.

24 Field et al., 1998. 'Weight Cycling, Weight Gain, and Risk of Hypertension in Women'. *American Journal of Epidemiology* 150 (6), 573–9.

25 Stevens et al., 2012. 'Weight Cycling and Mortality in a Large Prospective US Study'. *American Journal of Epidemiology*, 175 (8), 785–92.

26 Auyeung et al., 2010. 'Survival in Older Men May Benefit From Being Slightly Overweight and Centrally Obese – A 5 Year Follow-up Study in 4,000 Older Adults Using DXA'. *The Journal of Gerontology Series A: Biological Sciences and Medical Sciences*, 65 (1), 99–104.

Part Two – Inside Story: The Biology of Obesity

1 Dyson, F., 2007. 'Our Biotech Future'. *The New York Review of Books*. http://www. nybooks.com/articles/archives/2007/jul/19/our-biotech-future/ accessed September 2014.

Chapter 4 – Born to Become Obese?

1 Neena Modi, quoted in a Press Release issued by Imperial College London, September 2011.

2 Pond, C. M., 2003. 'Paracrine interactions of mammalian adipose tissue'. *Journal of Experimental Zoology, Part A. Comparative Experimental Biology*, 295 (1), 99–110; Elephant seals (2%), Fur Seals (6%) or even Harp seals (10.6%). From Kuzawa, K. W., 1998. 'Adipose Tissue in Human Infancy and Childhood: An Evolutionary Perspective'. *Yearbook of Physical Anthropology*, 41, 177–209; Nutritional Needs of Infants, Chapter 1 (Nutritional Needs of Infants), 15. This formula was calculated by the caloric RDI suggested for newborns, months 1–12. http://www.nal.usda.gov/ wicworks/Topics/FG/Chapter1_NutritionalNeeds.pdf accessed 3 September 2014.

3 Schmelzle, H. R., & Fusch, C., 2002. 'Body fat in neonates and young infants: validation of skinfold thickness versus dual-energy X-ray absorptiometry'. *American Journal of Clinical Nutrition*, 76 (5), 1096–1100.

4 Oftedal, O. T., et al., 1989. 'Effects of suckling and the post suckling fast on weights of the body and internal organs of harp and hooded seal pups'. *Biology Neonate*, 56 (5), 283–300; Lewis, D. S., et al., 1983. 'Preweaning nutrition and fat development in baboons'. *Journal of Nutrition*, 113 (11), 2253–9.

5 Schultz, A., 1969. *The Life of Primates.* New York: Universe Books.

6 Sinclair, D., 1985. *Human Growth after Birth.* Oxford: Oxford University Press.

7 Alexander, G., 1975. 'Body temperature control in mammalian young'. *British Medical Bulletin*, 31 (1), 62–8.

8 Schaefer, O., 1977. 'Are Eskimos more or less obese than other Canadians? A comparison of skin fold thickness and ponderal index in Canadian Eskimos'. *American Journal of Clinical Nutrition* 30 (10), 1623–8; Shephard, R., 1991. *Body Composition in Biological Anthropology.* Cambridge: Cambridge University Press.

9 Stini, W., 1981. 'Body composition and nutrient reserves in evolutionary perspective'. *World Review of Nutrition and Dietetics*, 37, 55–83.

10 Max Westenhöfer's strange speculations were published in 1942 when Nazism was at its height. In his book, *Der Eigenweg des Menschen* (P. Beck, trans. The Unique Road to Man. Berlin: W. Mannstaedt & Co.), he wrote: 'Homo sapiens cannot be a close relative of primates, because the specialisations of primates are those typical of mammals, but those of humans are those of an animal closer to the root of mammalian evolution. In other words, humans have less specialised features, such as feet and skeleton and organs, which seem closer to the "original" mammalian design than to the tree-climbing primate. This early relative, the proto-mammal/human, he proposes, may have been an amphibian, such as a salamander. His discussion of the Aquatic Ape Theory, or the aquatic phase in human evolution, as he terms it, is therefore used to support this amphibian ancestor of humans.' For an account of his work and his theory go to http://www.riverapes.com/AAH/Westnfr/WestnHfr.htm

11 Hardy, A., 1960. 'Was man more aquatic in the past?' *The New Scientist*, 17 March, 642–5.

12 Morgan, E., Aquatic Ape Theory, see: www.primitivism.com/aquatic-ape.htm accessed 15 August 2013.

13 One of the most widely used equations for calculating the brain size of placental animals is: $\log_{10}E = 0.76\log_{10}P + 1.77$ (Martin, R. D., 1983. 'Human Brain Evolution in an Ecological Context', 52nd James Arthur Lecture on the Evolution of the Human Brain, American Museum of Natural History, New York). Here E = brain mass in milligrams and P = body mass in grams. This equation gives an average human brain that is 4.6 times the size expected for the average mammal.

14 Leonard, W., & Robertson, M., 1994. 'Evolutionary perspectives on human nutrition: The influence of brain and body size on diet and metabolism'. *American Journal of Human Biology*. 6 (1), 77–88.

15 Aschoff, J., Günther, B. & Kramer, K., 1971. *Energiehaushalt und Temperaturregulation*. Munich: Urban and Schwarzenberg.

16 Martin, ' Human Brain Evolution'; Foley, R. A. & Lee, P. C., 1991. 'Ecology and energetics of encephalisation in hominid evolution'. *Philosophical Transactions of the Royal Society of London* 334 (1270), 223–32; McNab, B. K. & Eisenberg, J. F., 1989. 'Brain size and its relation to the rate of metabolism in mammals'. *The American Naturalist*, 133 (2), 157–67.

17 Schofield, C., 1985. 'An annotated bibliography of source material for basal metabolic rate data'. *Human Nutrition: Clinical Nutrition* 39C (suppl I), 42–91.

18 Aiello, L. C. & Wheeler, P., 1995. 'The Expensive Tissue Hypothesis: the brain and the digestive system in human and primate evolution'. *Current Anthropology* 36 (2), 199–221.

19 Ibid.

20 Expected values for brain mass organs was calculated, from data provided by Stephan, H., Frahm, H. & Baron, G., 1981. 'New and revised data on volume of brain structures in insectivores and primates'. *Folia Primatologica*, 35 (1), 1–29. Using the formula: Brain mass: $\log_{10}BW = 0.72 * \log_{10}M + 1.35$ ($N = 26$, $r = 0.98$). Gut mass from Chivers, D. J. & Haldik, C. M., 1980. 'Morphology of the gastrointestinal tract in primates: comparisons with other mammals in relation to diet'. *Journal of Morphology*, 166 (3), 337–86. Using the formula: $\text{Log}_{10}GM = 0.853 * \text{Log}_{10}M - 1.271$ ($N = 22$, $r = 0.96$). Where GM = gut mass in kilograms; W = other organ mass in grams; M = body mass in kilograms; n = number of individual animals; N = number of species; r = product moment correction coefficient. Diagram and notes reproduced from Aiello and Wheeler, 1995, with permission.

21 Aiello, L., 1997. 'Brains and guts in human evolution: The Expensive Tissue Hypothesis'. *Brazilian Journal of Genetics*. 20 (1), http://dx.doi.org/10.1590/S0100-84551997000100023

22 There is considerable variation in the size of the small intestine, where most of the digestion takes place, which range from 15 feet (4.6 m) to 32 feet (9.8 m).

23 Allard, Alexander Jr., 1972. *The Human Imperative*. New York: Columbia University Press.

24 Pilbeam, D. & Gould, S. J., 1974. 'Size and scaling in human evolution'. *Science*, 186 (4167), 892–901.

25 Dunbar, R. I., 2009. 'The social brain hypothesis and its implications for social evolution'. *Annals of Human Biology*, 36 (5), 562–72.

26 Koojiman, S. A. L. M., 1986. 'Energy budgets can explain body size relations'. *Journal of Theoretical Biology*, 121 (3), 269–82.

27 Lindlahr, V., 1942. *You Are What You Eat: how to win and keep health with diet.* Newcastle: National Nutrition Society.

28 http://www.reddit.com/r/askscience/comments/17txer/ did_humans_begin_cooking_ food_due_to_preference/ accessed 25 July 2014]; Carmody, R. N. & Wrangham, R. W., 2009. 'The energetic significance of cooking'. *Journal of Human Evolution*, 57 (4), 379–91.

29 Wrangham, R., 2009. *Catching Fire: How Cooking Made Us Human.* London: Profile Books.

30 Goren-Inbar, N., et al., 2004. 'Evidence of Hominin Control of Fire at Gesher Benot Ya'aqov Israel'. *Science*, 304 (30), 725–7.

31 Neel, J. V., 1962. 'Diabetes Mellitus: A "Thrifty" Genotype Rendered Detrimental by Progress'. *American Journal of Human Genetics* 14 (4), 353–62.

32 Ibid.

33 Eriksson, J. G., et al., 2003. 'Early adiposity rebound in childhood and risk of Type 2 diabetes in adult life'. *Diabetologia*, 46 (2), 190–4.

34 Barker, D. J., 2004. 'The developmental origins of adult disease'. *Journal of the American College of Nutrition*, 23 (6 Suppl), 588S–595S; Barker, D. J., 1994. *Mothers, Babies and Disease in Later Life.* London: BMJ Publishing Group; Hales, C. N., & Barker, D. J., 1992. 'Type 2 (non-insulin-dependent) diabetes mellitus: the thrifty phenotype hypothesis'. *Diabetologia*, 35 (7), 595–601.

35 Paneth, N. & Susser, M. 'Early origins of coronary heart disease (the "Barker hypothesis")' *British Medical Journal* editorial, published 18 February 1995. It should be noted that while the Barker hypothesis is widely accepted, it has yet to be completely proved.

36 Hales & Barker, 'The thrifty phenotype'.

37 Knowler, W. C., et al., 1983. 'Diabetes mellitus in the Pima Indians: genetic and evolutionary considerations'. *American Journal of Physical Anthropology*, 62 (1), 107–14; Swinburn, B., 1995. 'The thrifty genotype hypothesis: concepts and evidence after 30 years'. *Asia Pacific Journal of Clinical Nutrition*, 4, 337–8.

38 Gurney, R., 1936. 'The Hereditary Factor in Obesity'. *Archives of Internal Medicine*, 57 (3), 557–61.

39 Logue, A. W., 1986. *The Psychology of Eating and Drinking.* New York: W.H. Freeman and Company, 168.

40 Modi, N., et al., 2011. 'The Influence of Maternal Body Mass Index on Infant Adiposity and Hepatic Lipid Content'. *Pediatric Research* 70 (3), 287–91.

41 Ibid.

42 Dr Sean Kelly, personal communication, 2014.

Chapter 5 – The Secret Life of Fat

1 Quoted from http://www.webmd.com/diet/features/the-truth-about-fat

2 Ahima, R. S., 2006. 'Adipose tissue as an endocrine organ'. *Obesity (Silver Spring)*, 14 (Suppl. 5), 242S–249S; Ahima, R. S. & Flier, J. S., 2000. 'Adipose tissue as an endocrine organ'. *Trends Endocrinology & Metabolism*, 11 (8), 327–32.

3 Siiteri, P. K., 1987. 'Adipose tissue as a source of hormones'. *American Journal of Clinical Nutrition*, 45 (Suppl. 1), 277–82.

4 Wozniak, S. E., et al., 2009. 'Adipose tissue: the new endocrine organ?' *Digestive Diseases and Sciences*, 54 (9), 1847–56.

5 They are named after a 22-year-old German anatomist, Paul Langerhans, who discovered them in 1869. Langerhans P., 1869. 'Beitrage zur mikroscopischen anatomie der bauchspeichel druse'. Inaugural dissertation. Berlin: Gustav Langerhans.

6 Deng, T., et al., 2013. 'Class II Major Histocompatibility Complex Plays an Essential Role in Obesity-Induced Adipose Inflammation'. *Cell Metabolism*, 17 (3), 411–22.

7 Gandotra, S. et al., 2011. 'Perilipin deficiency and autosomal dominant partial lipodystrophy'. *New England Journal of Medicine*, 364 (8), 740–89.

8 Steppan, C. M., et al., 2001. 'The hormone resistin links obesity to diabetes'. *Nature* 409 (6818), 307–12.

9 McTernan, C. L., et al., 2002. 'Resistin, central obesity, and type 2 diabetes'. *The Lancet* 359 (9300), 46–7.

10 Ohashi, K., et al., 2006. 'Adiponectin replenishment ameliorates obesity-related hypertension'. *Hypertension.* 47 (6), 1108–16.

11 Wang, Z. V. & Scherer, P. E., 2008. 'Adiponectin, Cardiovascular Function, and Hypertension'. *Hypertension* 51, 8–14.

12 Ohashi, et al. 'Adiponectin replenishment'.

13 Qi, Y., et al., 2004. 'Adiponectin acts in the brain to decrease body weight'. *Nature Medicine*, 10 (5), 524–9.

14 Quotation from 'Key Hormone Protects Obese Mice from Diabetes', 28 August 2007. *Science Daily*, http://www.sciencedaily.com/releases/2007/08/070823201215.htm

15 Khamsi, R., 2007. 'World's fattest mouse appears immune to diabetes'. *New*

Scientist Health, 23 August, http://www.newscientist.com/article/dn12530-worlds-fattest-mouse-appears-immune-to-diabetes-.html#.Uv-wfPZSTGI

16 Neil, U., 2010. 'Leaping for leptin' *The Journal of Clinical Investigation* 120 (10), 3413–18

17 Masharani, U. & Gitelman, S., 'Pancreatic Hormones & Diabetes Mellitus'. In D. G. Gardner & D. Shoback, eds. *Greenspan's Basic & Clinical Endocrinology*. 9 ed. New York: McGraw-Hill Medical, 573–657.

18 Hummel, K. P., Dickie, M. M. & Coleman, D. L., 1966. 'Diabetes, a new mutation in the mouse'. *Science*. 153 (740), 1127–8.

19 Coleman, D. L., 2010. 'A historical perspective on leptin'. *Nature Medicine*, 16 (10), ix–xi.

20 Zhang, Y., et al., 1994. 'Positional cloning of the mouse obese gene and its human homologue'. *Nature*, 372 (6505), 425–32.

21 100 Million Dieters, $20 Billion: The Weight-Loss Industry by the Numbers. ABC News. 8 May 2012. http://abcnews.go.com/Health/100-million-dieters-20-billion-weight-loss-industry/story?id=16297197 accessed 4 September 2014.

22 Jeffrey Friedman, personal communication, 17 August 2014.

23 Neil, U., 2013. A Conversation with Jeffrey Friedman in *Conversations with Giants in Medicine*. The Journal of Clinical Investigation 123 (2), 529–30.

24 Ibid.

25 Jeffrey Friedman, personal communication, 10 August 2014.

26 Ibid.

27 Neil, A Conversation with Jeffrey Friedman.

28 Friedman, J. M. & Halaas, J. L., 1998. 'Leptin and the regulation of body weight in mammals'. *Nature* 395 (6704), 763–70.

29 Stetka, B. S. & Volkow, N. D., 2013. 'Can Obesity be an Addiction?' *Medscape*, http://www.medscape.com/viewarticle/807684 accessed 20 August 2014.

30 Cock, T. A. & Auwerx, J., 2003. 'Leptin: cutting the fat off the bone'. *The Lancet*, 362 (9395), 1572–4; Margetic, S., et al., 2002. 'Leptin: a review of its peripheral actions and interactions'. *International Journal of Obesity*, 26 (11), 1407–33.

31 Montague, C. T., et al., 1997. 'Congenital leptin deficiency is associated with severe early-onset obesity in humans'. *Nature*, 387 (6636), 903–8.

32 Heymsfield, S. B., et al., 1999. 'Recombinant leptin for weight loss in obese and lean adults: a randomized, controlled, dose-escalation trial'. *Journal of the American Medical Association* 282 (16), 1568–75.

33 Neil, A Conversation with Jeffrey Friedman.

34 Roth, J. D., et al., 2008. 'Leptin responsiveness restored by amylin agonism in diet-induced obesity: evidence from nonclinical and clinical studies'. *Proceedings of the National Academy of Science, USA.* 105 (20), 7257–62.

35 Cannon, B. & Nedergaard, J., 2004. 'Brown adipose tissue: function and physiological significance'. *Physiological Review,* 84 (1), 277–359.

36 Cypess, A. M., et al., 2009. 'Identification and Importance of Brown Adipose Tissue in Adult Humans'. *New England Journal of Medicine,* 360 (15), 1509–17.

37 Quoted in a news release dated 8 April 2009 from the Joslin Diabetes Center, Boston. 'Joslin Study Identifies "Good" Energy Burning Fat in Lean Adults. Finding May Lead to Treatments for Obesity, Diabetes.'

38 Virtanen, K.A., Lidell, M.E., Orava, J., et al., 2009 Functional brown adipose tissue in healthy adults. *New England Journal of Medicine,* 360 (15), 1518–25.

39 Chondronikola, M., Volpi, E., et al., 2014. Brown Adipose Tissue Improves Whole Body Glucose Homeostasis and Insulin Sensitivity in Humans. *Diabetes.* 63(12), 4089–99.

40 http://www.medscape.com/viewarticle/828978#vp_2

Chapter 6 – Gut Reactions

1 Hung, I. W. & Labroo, A. A., 2011. 'From Firm Muscles to Firm Willpower: Understanding the Role of Embodied Cognition in Self-Regulation'. *Journal of Consumer Research,* 37 (6), 1046–63.

2 De Vrieze, J., 2013. 'The Promise of Poop'. *Science,* 341 (6149), 954–7.

3 Jarvis, W. R., et al., 2009. 'National point prevalence of Clostridium difficile in US health care facility inpatients'. *American Journal of Infection Control* 37 (4), 263–70.

4 De Vrieze, 'Promise of Poop'.

5 Eiseman, B., et al., 1958. 'Fecal enema as an adjunct in the treatment of pseudomembranous enterocolitis'. *Surgery* 44 (5), 854–9

6 http://www.hopkinschildrens.org/Hope-for-Cure-of-Childhood-Diarrhea-Comes-Straight-from-the-Gut.aspx accessed 12 December 2014.

7 Faeces constitutes a level 2 biohazard, so before any can be implanted in a sick patient the donor has to be screened for such communicable diseases, as HIV, hepatitis and other disease-causing germs. The stools are diluted with saline solution or 4 per cent milk before being blended and fed into the patient's digestive system.

This can be done in a number of ways, including via the nasal passage (nasogastric) or nasoduodenal tubes, through a colonoscope or an enema. Stools are typically donated by spouses or other family members, but some doctors have used unrelated donors. Reference: Bakken, J. S., et al., 2011. 'Treating Clostridium difficile Infection With Fecal Microbiota Transplantation'. *Clinical Gastroenterology and Hepatology* 9 (12), 1044–9.

8 Borody, T. J., et al., 2004. 'Bacteriotherapy using faecal flora: toying with human motions'. *Journal of Clinical Gastroenterology*, 38 (6), 475–83.

9 Ibid.

10 Ibid.

11 Quoted from http://www.ucmp.berkeley.edu/history/leeuwenhoek.html

12 Dubner, S. J., 2011. 'The Power of Poop'. *Freakonomics.com*

13 Luiggi, 'Same poop, different gut'; de Vos, W. M. & de Vos, E. A., 2012. 'Role of the intestinal microbiome health and disease: from correlation to causation'. *Nutrition Reviews*, 70 (Suppl. 1), S45–56; Reid, G., et al., 2011. 'Microbiota restoration: natural and supplemented recovery of human microbial communities'. *National Review of Microbiology*, 9 (1), 27–38; Lozupone, C. A., et al., 2012. 'Diversity, stability and resilience of the human gut microbiota'. *Nature* 489 (7415), 220–30.

14 http://nih.gov/news/health/jun2012/nhgri-13.htm.

15 Ackerman, J., 2012. 'The ultimate social network'. *Scientific American*, 306 (6), 21–7.

16 O'Hara, A. M. & Shanahan, F., 2006. 'The gut flora as a forgotten organ'. *European Molecular Biology Organisation (EMBO)*, 7 (7), 688–93.

17 Costello, E. K., et al., 2012. 'The Application of Ecological Theory Toward an Understanding of the Human Microbiome'. *Science*, 336 (8), 1255–62.

18 http://science.time.com/2013/08/29/you-are-your-bacteria-how-the-gut-microbiome-influences-health

19 Velasquez-Manoff, M., 2013. *An Epidemic of Absence: A New Way of Understanding Allergies and Autoimmune Diseases.* New York, Simon & Schuster.

20 Dominguez-Belloa, M. G., et al., 2010. 'Delivery mode shapes the acquisition and structure of the initial microbiota across multiple body habitats in newborns'. *Proceedings of the National Academy of Science.* 107 (26), 11971–5; Mändar, R. & Mikelsaar, M., 1996. 'Transmission of mother's microflora to the newborn at birth'. *Biology of the Neonate*, 69 (1), 30–5.

21 Ursell, L. K., et al., 2012. 'Defining the human microbiome'. *Nutrition Reviews*, 70 (Suppl. 1), S38–44.

22 Palmer, C., et al., 2007. 'Development of the human infant intestinal microbiota'. *Public Library of Science Biology*, 5 (7), e177.

23 Gura, T., 2014. 'Nature's first functional food'. *Science* 345 (6198), 747–9.

24 Ibid.

25 Quoted in Ibid.

26 Khanna, S. & Tosh, P. K., 2014. 'A clinician's primer on the role of the microbiome in human health and disease'. *Mayo Clinical Proceedings*, 89 (1), 107–14.

27 Tilg, H., & Kaser, A., 2011. 'Gut microbiome, obesity, and metabolic dysfunction'. *Journal of Clinical Investigation*, 121 (6), 2126–32.

28 https://microbewiki.kenyon.edu/index.php/Firmicutes_and_Obesity; http://www.mf.uni-mb.si/mf/instituti/IPweb/html/DiGioiaD%20Introduction%20to%20intestinal%20microbiota.pdf accessed 19 November 2014

29 Boyanova, L., ed. 2011. *Helicobacter pylori.* Norfolk: Caister Academic Press.

30 Ley is quoted from website of Cornell University. http://micro.cornell.edu/people/ruth-ley, , downloaded March 2014.

31 Turnbaugh, P. J., et al., 2009. 'The effect of diet on the human gut microbiome: a metagenomic analysis in humanized gnotobiotic mice'. *Science of Translational Medicine*, 1 (6), 6–14.

32 Tilg, H. & Moschen, A. R., 2014. 'Microbiota and diabetes: an evolving relationship'. *Gut*, 63 (9), 1513–21.

33 Blaser, M. J., 2005. 'Global warming and the human stomach: Microecology follows macroecology'. *Transactions of the American Clinical and Climatological Association*, 116, 65–76.

34 Ibid.

35 Ibid.

36 François, F., et al., 2011. 'The effect of H. pylori eradication on meal-associated changes in plasma ghrelin and leptin'. *BMC Gastroenterology* 11 (2011), 37.

37 Kalat, J. W., 1998. *Biological Psychology*, 6th Edition, Wadsworth: Cengage Learning, USA.

38 Cryan, J. F. & Dinan, T. G., 2012. 'Mind-altering microorganisms: the impact of the gut microbiota on brain and behaviour'. *National Review of Neuroscience*, 13 (10), 701–12.

39 Clarke, G., et al., 2013. 'The microbiome-gut-brain axis during early life regulates the hippocampal serotonergic system in a sex-dependent manner'. *Molecular Psychiatry*, 18 (6), 666–73.

40 Turnbaugh, P. J., Ley, R. E., et al. , 2006. 'An obesity-associated gut microbiome with increased capacity for energy harvest'. *Nature,* 444 (11). 1027-1031.

41 Wisknewsky, J., Doré, J., Clement, K., 2012. The importance of gut microbiota after bariatric surgery. *Nature Reviews Gastroenterology & Hepatology.* 9 (11) pp 590-598; Bäckhed, F., et al., 2004. 'The gut microbiota as an environmental factor that regulates fat storage'. *Proceedings of the National Academy of Science,* 101 (44), 15718–23; Bäckhed, F., et al., 2005. 'Host-bacterial mutualism in the human intestine'. *Science* 307 (5717), 1915–1920.

42 Tremaroli, V. & Bäckhed, F., 2012. 'Functional interactions between the gut microbiota and host metabolism'. *Nature,* 489 (7415), 242–9.

43 Turnbaugh, P. J., & Gordon, J. I., 2009. 'The core gut microbiome, energy balance and obesity'. *Journal of Physiology,* 587 (17), 4153–8; Vrieze, A., et al., 2012. 'Transfer of Intestinal Microbiota From Lean Donors Increases Insulin Sensitivity in Individuals With Metabolic Syndrome'. *Gastroenterology* 143 (4), 913–16.

44 Ibid.

45 Li, J., et al., 2007. 'Gene function prediction based on genomic context clustering and discriminative learning: an application to bacteriophages'. *BMC Bioinformatics,* 8 (Suppl. 4).

Part Two – All in the Mind? Obesity and Your Brain

1 Smail, D. L., 2008. *On Deep History and the Brain.* Berkeley, CA: University of California Press, 140.

Chapter 7 – How Your Brain Can Make You Obese

1 F. Adams, trans. *On the Sacred Disease* by Hippocrates. Internet Classics Archive by Daniel C. Stevenson, downloaded from http://classics.mit.edu/Hippocrates/sacred.html

2 Nummenmaa L., et al., 2012. 'Dorsal striatum and its limbic connectivity mediate abnormal anticipatory reward processing in obesity'. *PloS One,* 7 (2), e31089.

3 Berthoud, H. R., Lenard, N. R. & Shin, A. C., 2011. 'Food reward, hyperphagia, and obesity'. *American Journal of Physiology.* 300 (6), R1266–77; Stice, E., et al., 2009. 'Relation of obesity to consummatory and anticipatory food reward'. *Physiology and Behavior,* 97 (5), 551–60.

4 Rothemund, Y., et al., 2007. 'Differential activation of the dorsal striatum by high-calorie visual food stimuli in obese individuals'. *Neuroimage,* 37 (2), 410–21.

5 Du, H., et al., 'Dietary energy density in relation to subsequent changes of weight and waist circumference in European men and women'. *PloS One*, 4 (4), e5339; Forouhi, N. G., et al., 2009. 'Dietary fat intake and subsequent weight change in adults: results from the European Prospective Investigation into Cancer and Nutrition cohorts'. *The American Journal of Clinical Nutrition*, 90 (6), 1632–41.

6 Nisbett, R. E., 1968. 'Determinants of food intake in obesity'. *Science*, 159 (820), 1254–5.

7 Ng, M., et al., 2014. 'Global, regional, and national prevalence of overweight and obesity in children and adults during 1980–2013: a systematic analysis for the Global Burden of Disease Study 2013'. *The Lancet*, 384 (9945), 766–81.

8 Petrovich, G. D., Holland, P. C. & Gallagher, M., 2005. 'Amygdalar and prefrontal pathways to the lateral hypothalamus are activated by a learned cue that stimulates eating'. *The Journal of Neuroscience*, 25 (36), 8295–302.

9 Jerison, H. J., 'Evolution of the Frontal Lobes'. In B. L. Miller & J. L. Cummings, eds. *The Human Frontal Lobes: Functions and Disorders*. New York: Guilford Press, 2007.

10 Stice, E., et al., 2010. 'Weight gain is associated with reduced striatal response to palatable food'. *The Journal of Neuroscience*, 30 (39), 13105–9.

11 Castellanos, E. H., et al., 2009. 'Obese adults have visual attention bias for food cue images: evidence for altered reward system function'. *International Journal of Obesity*, 33 (9), 1063–73; Stoeckel, L. E., et al., 2008. 'Widespread reward-system activation in obese women in response to pictures of high-calorie foods'. *Neuroimage*, 41 (2), 636–47; Stoeckel, L. E., et al., 2009. 'Effective connectivity of a reward network in obese women'. *Brain Research Bulletin*, 79 (6), 388–95.

12 Burke K. A., Franz T. M., Miller D. N., Schoenbaum G. The role of the orbitofrontal cortex in the pursuit of happiness and more specific rewards. *Nature*. 2008; 454(7202):340-344; Gottfried J. A., O'Doherty J., Dolan R. J. Encoding predictive reward value in human amygdala and orbitofrontal cortex. *Science*. 2003; 301(5636):1104-1107.

13 Sinha, R. & Jastreboff, A. M., 2013. 'Stress as a common risk factor for obesity and addiction'. *Biological Psychiatry*, 73 (9), 827–35; Parent, M. B., Darling, J. N. & Henderson, Y. O., 2014. 'Remembering to eat: hippocampal regulation of meal onset'. *American Journal of Physiology*, 306 (10), R701–13.

14 Wang, G. J., et al., 2006. 'Gastric stimulation in obese subjects activates the hippocampus and other regions involved in brain reward circuitry'. *Proceedings of the National Academy of Sciences, USA*, 103 (42), 15641–5.

15 Wicker, B., et al., 2003. 'Both of Us Disgusted in My Insula: the common neural basis of seeing and feeling disgust'. *Neuron*, 40 (3), 655–64.

16 von Deneen, K. M. & Liu, Y., 'Food Addiction, Obesity and Neuroimaging Addictions from Pathophysiology to Treatment'. In D. Belin, ed. *Addictions – From Pathophysiology to Treatment.* InTech, 259–90.

17 Bartoshuk, L. M., et al., 2006. 'Psychophysics of sweet and fat perception in obesity: problems, solutions and new perspectives'. *Philosophical Transactions of the Royal Society* London, 361 (1471), 1137–48; Volkow, N. D., 2013. 'The addictive dimensionality of obesity'. *Biological Psychiatry*, 73 (9), 811–18.

18 Berridge, K. C., et al., 2010. 'The tempted brain eats: pleasure and desire circuits in obesity and eating disorders'. *Brain Research*, 1350, 43–64; Pecina, S., Smith, K. S., Berridge, K. C., 2006. 'Hedonic hot spots in the brain'. *Neuroscientist*, 12 (6), 500–11.

19 Pecina, S., Smith, K. S., Berridge, K.C. 2006. 'Hedonic hot spots in the brain.' *Neuroscientist.* 12(6):500-511.

20 Pelchat, M., Johnson, A., Chan, R., Valdex, J., Ragland, J.D. 2004. 'Images of desire: food-craving activation during fMRI.' *Neuroimage.* 23:1486–93.

21 http://thebrain.mcgill.ca/flash/i/i_03/i_03_m/i_03_m_par/i_03_m_par_amphetamine.html

22 Wang, G. J., Volkow, N. D., Logan, J., et al. 2001. 'Brain dopamine and obesity.' *Lancet.* 357(9253):354-357

23 Pani, L. & Gessa, G. L., 1997. 'Evolution of the dopaminergic system and its relationships with the psychopathology of pleasure'. *International Journal of Clinical Pharmacology Research*, 17 (2–3), 55–8.

24 Lammers, G. J., et al., 1996. 'Spontaneous food choice in narcolepsy'. *Sleep*, 19 (1), 75–6; Schuld A., et al., 2000. 'Increased body-mass index in patients with narcolepsy'. *The Lancet*, 355 (9211), 1274–5.

25 Hara, J., et al., 2001. 'Genetic ablation of orexin neurons in mice results in narcolepsy, hypophagia, and obesity'. *Neuron*, 30 (2), 345–54; Kajiyama, S., et al., 2005. 'Spinal orexin-1 receptors mediate anti-hyperalgesic effects of intrathecally-administered orexins in diabetic neuropathic pain model rats'. *Brain Research*, 1044 (1), 76–86; Yamanaka A., Beuckmann C.T., Willie J.T., et al. Hypothalamic orexin neurons regulate arousal according to energy balance in mice. *Neuron.* 2003; 38(5):701–13.

26 Nummenmaa, L., Hirvonen, J., et al., 2012. 'Dorsal striatum and its limbic connectivity mediate abnormal reward processing in obesity'. *PLoS1.* 7(2): e31089.

27 Volkow, N. 2013. 'The addictive dimensionality of obesity'. *Biological Psychiatry*, 73, 811–818.

28 Wang, G.J., Volkow, N.D., Telang, F., Jayne, M., Ma, J. et al., 2004, 'Exposure to appetitive food stimuli markedly activates the human brain. *Neuroimage*, 21, 1790–7.

29 Geiger, B.M., 2008. 'Evidence for defective mesolimbic dopamine exocytosis in obesity-prone rats.' FASEB J. *Federation of the American Societies for Experimental Biology*, 22, 2740–6.

30 Anzman, S. L., Rollins, B. Y., & Birch, L. L., 2010. 'Parental influence on children's early eating environments and obesity risk: implications for prevention. *International Journal of Obesity*, 34(7), 1116–24.

31 Stice, E., 2008. 'Relation between obesity and blunted striatal response to food is moderated by TaqIA A1 allele.' *Science* 322, pp 449–52.

32 Burger, K. S., & Stice, E. (2011). 'Variability in reward responsivity and obesity: evidence from brain imaging studies'. *Current Drug Abuse Reviews*, 4(3), 182–9.

33 Volkow quote in Setka, B., 2013. 'Can obesity be an addiction?' http://www.medscape.com/viewarticle/807684

Chapter 8 – 'I heard an ice-cream call my name': Why Impulsive Eating Means Overeating

1 Kessler, D., 2009. *The End of Overeating: Taking Control of the Insatiable American Appetite*. Rodale, New York.

2 Lejuez, C. W., et al., 2003. 'Balloon Analogue Risk Task (BART) differentiates smokers and nonsmokers'. *Experimental and Clinical Psychopharmacology*, 11 (1), 26–33; Lejuez, C. W., et al., 2002. 'Evaluation of a behavioral measure of risk taking: the Balloon Analogue Risk Task (BART)'. *Journal of Experimental Psychology. Applied*, 8 (2), 75–84; Lejuez, C. W., et al., 2003. 'Evaluation of the Balloon Analogue Risk Task (BART) as a predictor of adolescent real-world risk-taking behaviours'. *Journal of Adolescence*, 26 (4), 475–9.

3 Hunt, M. K., et al., 2005. 'Construct validity of the Balloon Analog Risk Task (BART): associations with psychopathy and impulsivity'. *Assessment*, 12 (4), 416–28.

4 Lejuez, 'Balloon Analogue Risk Task (BART)'.

5 Herman, C. P. & Polivy, J., 1975. 'Anxiety, restraint and eating behavior'. *Journal of Abnormal Psychology*, 84 (6), 66–72; Herman, C. P. & Polivy, J., 1982. 'Weight change and dietary concern in the overweight: are they really independent?' *Appetite*, 3 (3), 280–1; Polivy, J., Herman, C. P. & Warsh, S., 1978. 'Internal and external components of emotionality in restrained and unrestrained eaters'. *Journal of Abnormal Psychology*, 87 (5), 497–504.

6 Atkins website, http://www.atkins.com/Science/Articles---Library/Carbohydrates/What-are-Net-Carbs-.aspx accessed September 2014.

7 Polivy, 'Internal and external components' http://www.spring.org.uk/2011/03/the-what-the-hell-effect.php

8   Herman, C. P. & Polivy, J., 1990. 'From dietary restraint to binge eating: attaching causes to effects'. *Appetite*, 14 (2), 123–5.

9   Joseph, R. J., et al., 2011. 'The neurocognitive connection between physical activity and eating behavior'. *Obesity Reviews*, 12 (10), 800–12.

10   Eaton, C. B., et al., 1995. 'Cross-sectional relationship between diet and physical activity in two southeastern New England communities'. *American Journal of Preventative Medicine.* 11 (4), 238–44.

11   Colcombe, S. J., et al., 2003. 'Aerobic fitness reduces brain tissue loss in aging humans'. *Journal of Gerontology: Medical Sciences*, 58A (2), 176– 180; Voss M. W., et al., 2010. 'Plasticity of brain networks in a randomized intervention trial of exercise training in older adults'. *Front Aging Neuroscience.* 26 (2), ii:32; Colcombe, S. J., et al., 2006. 'Aerobic exercise training increases brain volume in aging humans'. *Journal of Gerontology: Medical Sciences*, 61A (11), 1166–70.

12   Kramer, A. F., et al., 1999. 'Ageing, fitness, and neurocognitive function'. *Nature.* 400 (July), 418–19.

13   Allison, D. B., et al., 1993. 'Evidence of commingling in human eating behavior'. *Obesity Research*, 1 (5), 339–44; Nederkoorn, C., et al., 2006. 'Why obese children cannot resist food: The role of impulsivity'. *Eating Behavior*, 7 (4), 315–22; Guerrierri, R., 2007. 'How impulsiveness and variety influence food intake in a sample of healthy women'. *Appetite*, 48 (1), 119–22; Carrard, I., et al., 2012. 'Relations between pure dietary and dietary-negative affect subtypes and impulsivity and reinforcement sensitivity in binge eating individuals'. *Eating Behavior*, 13 (1), 13–19.

14   Beaver, J. D., et al., 2006. 'Individual differences in reward drive predict neural responses to images of food'. *Journal of Neuroscience*, 26 (19), 5160–6.

15   Berridge, K. C., et al., 2010. 'The tempted brain eats: pleasure and desire circuits in obesity and eating disorders'. *Brain Research*, 1350, 43–64.

16   Baumeister, R., 2012. 'Self-Control – the moral muscle'. *The Psychologist*, 25 (2), 112–15.

17   Baumeister, R. F., et al., 1998. 'Ego depletion: Is the active self a limited resource?' *Journal of Personality and Social Psychology*, 74 (5), 1252–6.

18   Baumeister, R. F., et al., 2008. 'Free will in consumer behavior: Self-control, ego depletion, and choice'. *Journal of Consumer Psychology*, 18 (2008), 4–13.

19   Baumeister, 'Self-Control'.

20   Ibid.

21   Ackerman, J. M., et al., 2009. 'You Wear Me Out: The vicarious depletion of self-control'. *Psychological Science*, 20 (3), 326–32.

22 Ibid.

23 Inzlicht, M. & Schmeichel, B. J., 2012. 'What is ego depletion? Toward a mechanistic revision of the resource model of self-control. *Perspectives on Psychological Science*, 7 (5) 450–63; Lowenstein, G., 1996. 'Out of control: visceral influences on behavior'. *Organizational Behavior and Human Decision Processes*, 65 (3), 272–92.

24 Dörner, D., 1996. *The Logic of Failure*. New York: Metropolitan Books, 28–34; Anderson, C., Platten, C. R., 2011. 'Sleep deprivation lowers inhibition and enhances impulsivity to negative stimuli'. *Behavioural Brain Research*, 217 (2), 463–6; Haghighatdoost, F., et al., 2012. 'Sleep deprivation is associated with lower diet quality indices and higher rate of general and central obesity among young female students in Iran'. *Nutrition*, 28 (11–12), 1146–50; Gangwisch, J. E., et al., 2005. 'Inadequate sleep as a risk factor for obesity: analyses of the NHANES I'. *Sleep*, 28 (10), 1289–96.

25 Schmid, S. M., et al., 2008. 'A single night of sleep deprivation increases ghrelin levels and feelings of hunger in normal-weight healthy men'. *Journal of Sleep Research*, 17, 331–4.

26 Conducted on behalf of *Secret Eaters*, Channel 4/Endemol. 12/13 December 2013.

27 Chaput, Jean-Philippe, personal communication, 20 August 2014.

28 Bruni, O., et al., 1996. 'The sleep disturbance scale for children (SDSC): childhood and adolescence'. *Journal of Sleep Research*, 5 (4), 252–61; Hawley, E., et al., 2004. 'Snoring and sleep disordered breathing in young children: subjective and objective correlates'. *Journal of Sleep Research*, 27 (1), 87–94; García, C. M., 2003. 'Neurobiología del transtorno de hiperactividad'. *Review of Neurology*, 36 (2003), 555–65.

29 Medeiros, M., et al., 2005. 'Sleep Disorders Are Associated With Impulsivity In School Children Aged 8 To 10 Years', *Arq Neuropsiquiatr* 63 (3-B), 761–5.

30 Koenig, S. M., 2001. 'Pulmonary complications of obesity'. *American Journal of Medical Science*, 321 (4), 249–79.

31 Jones, R. L. & Nzekwu, M. U., 2006. 'The effects of body mass index on lung volumes'. *Chest*. 130 (3), 827–33.

32 Lowe, M. & Eldredge, K., 1993. 'The Role of Impulsiveness in Normal and Disordered Eating'. In W. G. McCown, J. L. Johnson and M. B. Shure, eds. *The Impulsive Client, Theory, Research and Treatment*. Washington DC: American Psychological Association, 186–7.

33 Baumeister, R. F., Heatherton, T. F. & Tice, D. M., 1994. *Losing Control: How and why people fail at self-regulation*. San Diego, CA: Academic Press; Leitch, M. A., 2010. PhD thesis: 'Impulsivity and Overeating in Lean Females'. University of Sussex.

34 Shefer, G., Marcus, Y. & Stern, N., 2013. 'Is obesity a brain disease?' *Neuroscience and Biobehavioural Reviews*, 37 (10, Pt 2), 2489–503.

35 Pistell, P. J., et al., 2010. 'Cognitive impairment following high fat diet consumption is associated with brain inflammation'. *Journal of Neuroimmunology*, 219 (1–2), 25–32.

36 Letra, L., Santana, I. & Seica, R., 2014. 'Obesity as a risk factor for Alzheimer's disease: the role of adipocytokines'. *Metab Brain Dis.*, 29 (3), 563–8.

37 Craft, S., Cholerton, B. & Baker, L. D., 2013. 'Insulin and Alzheimer's disease: untangling the web'. *Journal of Alzheimer's Disease*, 33 (Suppl 1), S263–75; Parton, L. E., et al., 2007. 'Glucose sensing by POMC neurons regulates glucose homeostasis and is impaired in obesity'. *Nature*, 449 (7159), 228–32.

Chapter 9 – Feeding Our Feelings: The Power of Emotional Eating

1 Louis, C.K., *Chewed Up*, stand-up routine, 2008.

2 Study funded by Channel 4 as part of a *Secret Eaters* programme, transmitted in 2014.

3 Cornil, Y. & Chandon, P., 2013. 'From Fan to Fat? Vicarious Losing Increases Unhealthy Eating, but Self-Affirmation Is an Effective Remedy'. *Psychological Science*, 24 (10), 1936–46.

4 Quote taken from Kessler, D. A., 2009. *The End of Overeating*. London: Penguin Books, 150. Dr. Koob's research bridges the gap in understanding compulsive habits regarding consumption of food and drugs, and has been featured in David Kessler's ground breaking book *The End of Overeating*.

5 Chua, J. L., Touyz, S. & Hill, A. J., 2004. 'Negative mood-induced overeating in obese binge eaters: an experimental study'. *International Journal of Obesity and Related Metabolic Disorders*, 28 (4), 606–10.

6 Erlanson-Albertsson, C., 2005. 'How palatable food disrupts appetite regulation'. *Basic Clinical Pharmacology Toxicology*, 97 (2), 61–73.

7 Olson, J., 2014. *Counting Calories: A True Story From an Average Jane Who Lost Over 120 Pounds in Less than Six months*. Available from Amazon as a Kindle download.

8 2003. 'How America eats'. *Bon Appétit*. http://www.epicurious.com/bonappetit / features/survey2003

9 Witherly, S., 2007. *Why Humans Like Junk Food: The Inside Story on Why You Like Your Favorite Foods, the Cuisine Secrets of Top Chefs, and How to Improve Your Own Cooking*. Lincoln: Iuniverse Press.

10 Ibid.

11 Ibid.

12 Roxby, P., 2013. 'Chocolate craving comes from total sensory pleasure'. *BBC News Health.* http://www.bbc.com/news/health-23449795

13 2014. 'Worldwide Consumption Rates'. *The World Atlas of Chocolate.* http://www.sfu.ca/geog351fall03/groups-webpages/gp8/consum/consum.html accessed 5 June 2014.

14 Ibid.

15 Nieburg, O., 2013. 'Interactive Map: Top 20 chocolate-consuming nations of 2012'. Confectionery News.com. http://www.confectionerynews.com/Markets/Interactive-Map-Top-20-chocolate-consuming-nations-of-2012 accessed 5 June 2013; 2012. 'Who consumes the most chocolate'. *The CNN Freedom Project, Ending Modern Day Slavery.* http://thecnnfreedomproject.blogs.cnn.com/2012/01/17/who-consumes-the-most-chocolate accessed 5 June 2013.

16 Piomelli, D., 1996. 'Brain cannabinoids in chocolate'. *Nature,* 382 (August), 677–8.

17 Herraiz, T., 2000. 'Tetrahydro-beta-carbolines, Potential Neuroactive Alkaloids, in Chocolate and Cocoa'. *Journal of Agriculture Food Chemisty,* 48 (10), 4900–4.

18 Parker, G. & Crawford, J., 2007. 'Chocolate craving when depressed: a personality marker'. *The British Journal of Psychiatry,* 205 (2), 351–2.

19 Stice, E., et al., 2008. 'Relation of reward from food intake and anticipated food intake to obesity: A functional magnetic resonance imaging study'. *Journal of Abnormal Psychology,* 117 (4), 924–35.

20 Wang, G. J., et al., 2001. 'Brain dopamine and obesity'. *The Lancet,* 357 (9253), 354–7.

21 Avena, N. M., 2010. 'The study of food addiction using animal models of binge eating'. *Appetite,* 55 (2010), 734–7.

22 Avena, N. M., Long, K. A. & Hoebel, B. G., 2005. 'Sugar-dependent rats show enhanced responding for sugar after abstinence: evidence of a sugar deprivation effect'. *Physiology & Behavior,* 84 (3), 359–62.

23 Avena, 'Sugar-dependent rats', 735.

24 Avena, personal communication, 15 May 2014.

25 Colantuoni, C., et al., 2002. 'Evidence that intermittent, excessive sugar intake causes endogenous opioid dependence'. *Obesity Research,* 10 (6), 478–88.

26 Hoebel, B. G., et al., 2009. 'Natural addiction: a behavioral and circuit model based on sugar addiction in rats'. *Journal of Addiction Medicine*, 3 (1), 33–41.

27 Guertin, T. L., 1999. 'Eating behavior of bulimics, self-identified binge eaters, and non-eating-disordered individuals: what differentiates these populations?' *Clinical Psychological Review*, 19 (1), 1–23.

28 Lemmens, S. G., et al., 2011. 'Stress augments food "wanting" and energy intake in visceral overweight subjects in the absence of hunger'. *Physiology & Behavior*, 103 (2), 157–63.

29 Sproesser, G., Schupp, H. T. & Renner, B., 2013. 'The Bright Side of Stress-Induced Eating: Eating More When Stressed but Less When Pleased'. *Psychological Science*, published online 28 October.

30 Ibid.

31 Atkinson, F. S., Foster-Powell, K. & Brand-Miller, J. C., 2008. 'International Tables of Glycemic Index and Glycemic Load Values'. *Diabetes Care*, 31 (12), 2281–3.

32 Steward, H. L., Andrews, S. S, & Bethea, M. C., 2003. *The New Sugar Busters! Cut Sugar to Trim Fat*. New York: Ballantine Books, 69.

33 Benedict, C., et al., 2012. 'Impaired insulin sensitivity as indexed by the HOMA score is associated with deficits in verbal fluency and temporal lobe gray matter volume in the elderly'. *Diabetes Care*, 35 (3), 488–94.

34 Fotuhi, M., Do, D., Jack, C., 2012. 'Modifiable factors that alter the size of the hippocampus with ageing'. *Nature Reviews. Neurology*, 8 (4), 189–202; Hallschmid, M., et al. 2008. 'Obese men respond to cognitive but not to catabolic brain insulin signaling'. *International Journal of Obesity*, 32 (2), 275–82.

Part Four – Our Obesogenic World

1 Berridge, K., personal communication, 17 July 2014.

Chapter 10 – Caution: Food Cues at Work

1 Gordon Shepherd, quoted in Kessler, D., 2009. *The End of Overeating*. London: Penguin Books.

2 Berridge, K., personal communication, 15 July 2014.

3 Weingarten, H. P., 1983. 'Conditioned cues elicit feeding in sated rats: a role for learning in meal initiation'. *Science* 220 (4595), 431–3. Animal models offer insight into biologically ingrained behaviours. They provide paradigms uncomplicated by cultural or social norms and, while animals do not have the same cognitive complexity as humans, they provide the most robust platform for investigating the biological antecedents to eating. Scientists can manipulate contexts, social situations, food choice, and even compare the rewarding effects of sugar to class A drugs, like cocaine. Something which, for obvious ethical reasons, would not be possible with humans. While animals provide one tool for investigation, increasing emphasis has been put on computer modelling and digital replicas of biological phenomena. Moreover, we have also started to use tools like functional Magnetic Resonance Imaging, which allows a direct view into the way the human and animal brains behave.

4 Channel 4, 2012. *Secret Eaters*. Based on experimental protocol established by Brian Wansink at Cornell's Food and Brand Lab; Neal, D., et al., 2011. 'The Pull of the Past: When Do Habits Persist Despite Conflict With Motives?' *Personality and Social Psychology*, 37 (11), 1428–37.

5 Cornell, C. E., Rodin, J. & Weingarten, H., 1989. 'Stimulus-induced eating when satiated'. *Physiology & Behavior*, 45 (4), 695–704.

6 Berridge, personal communication, 15 July 2014.

7 Lewis, D., 2013. *The Brain Sell: When Science Meets Shopping*. London: Nicholas Brealey. Newsday described the technique as 'the most alarming invention since the atomic bomb,' while *Saturday Review* editor Norman Cousins urged the authorities 'to take this invention and everything connected to it and attach it to the centre of the next nuclear explosive scheduled for testing.' Cousins, N., 1957, 'Smudging the Subconscious', *Saturday Review*, 5 October, pp. 20–40.

8 Garfield, B., 2000. '"Subliminal" seduction and other urban myths'. *Advertising Age*, 18 September. Writing in 1981, John O' Toole, Chairman of Foote, Cone & Belding Communications, Inc., of one of the world's largest advertising agencies, flatly denied any such technique had ever been used and added: 'It is demeaning to assume that the human mind is so easily controlled that anyone can be made to act against his will or better judgement by peremptory commands he doesn't realise are present.' O'Toole, J., 1981. *The Trouble with Advertising*. New York: Chelsea House, 16.

9 Winkielman, P., Berridge, K. C. & Wilbarger, J. L., 2005. 'Unconscious affective reactions to masked happy versus angry faces influence consumption behavior and judgments of value'. *Personality & Social Psychology Bulletin*, 31 (1), 121–35.

10 Karremans, J. C., Stroebe, W. & Claus, J., 2006. 'Beyond Vicary's fantasies: The impact of subliminal priming and brand choice'. *Journal of Experimental Social Psychology*, 42 (6), 792–8.

11 Bálint, R., 1907. 'A nézés lelki bénulása, optikai ataxia, a figyelem térbeli zavara' (Psychic paralysis of gaze, optic ataxia, and disturbance of spatial attention). *Orvosi Hetilap* (Medicine Weekly) 1 (1907), 209–36. Cited in Husain

M. & Stein, J., 1988. 'Resö Bálint and his most celebrated case'. *Archives of Neurology*, 45 (1), 89–93.

12 Papies, E. K. & Hamstra, P., 2010. 'Goal Priming and Eating Behavior: Enhancing Self-Regulation by Environmental Cues'. *Health Psychology*, 29 (4), 384–8.

13 Kessler, *End of Overeating*.

14 Stoeckel, L. E., et al., 2009. 'Effective connectivity of a reward network in obese women'. *Brain Research Bulletin*, 79(6), 388–95.

15 Berridge, personal communication, 2014.

16 Berridge, K., 1996. 'Food reward: Brain substrates of wanting and liking'. *Neuroscience and Behavioral Reviews*, 20 (1), 1–25; Berridge, K. C., 2009. 'Wanting and Liking: Observations from the Neuroscience and Psychology Laboratory'. *Inquiry*, 52 (4), 378; Berridge, K. C., 2004. 'Motivation concepts in behavioral neuroscience'. *Physiology & Behavior*, 81 (2), 179–209; Berthoud, H. R., 2007. 'Interactions between the "cognitive" and "metabolic" brain in the control of food intake'. *Physiology & Behavior*. 91 (5), 486–98.

17 Kelley, A. E., et al., 2002. 'Opioid modulation of taste hedonics within the ventral striatum'. *Physiology & Behavior*, 76 (3), 365–77; Zhang, N., et al., 2000. 'A mutation in the Lunatic fringe gene suppresses the effects of a Jagged2 mutation on inner hair cell development in the cochlea'. *Current Biology*, 10 (11), 659–62.

18 Berridge, K. C. & Robinson, T. E., 1998. 'What is the role of dopamine in reward: hedonic impact, reward learning, or incentive salience?' *Brain Research Revue*, 28 (3), 309–69.

19 Robinson, T. E. & Berridge, K. C., 2003. 'Addiction'. *Annual Review of Psychology*. 54 (2003), 25–53; Cannon, C. M. & Palmiter, R. D., 2003. 'Reward without dopamine'. *The Journal of Neuroscience*, 23 (34), 10827–31.

20 Avena, N. M., Rada, P. & Hoebel, B. G., 2008. 'Evidence for sugar addiction: behavioral and neurochemical effects of intermittent, excessive sugar intake'. *Neuroscience and Biobehavioral Reviews*, 32 (1), 20–39.

21 At the time of writing *Fat Planet*, pharmacological intervention for obesity was limited. New developments of products such as Contrave's Qsymia, which is a combination drug of naltrexone (long-acting naloxone) and bupropion (antidepressant) is being considered by the FDA for treatment of obesity; http://www.forbes.com/sites/larryhusten/2014/09/10/fda-approves-contrave-weight-loss-drug-from-orexigen-and-takeda/

Chapter 11 – See Food – Eat Food: The Cues That Elicit Overeating

1 Kilbourne, J., 2004. 'The More You Subtract, the More You Add: Cutting Girls Down to Size'. In Kasser, T. & Kanner, A.D., eds. *Psychology and Consumer Culture.* Washington DC: American Psychological Association, 252.

2 'Restaurant Ad Spend Rises 4.4%; McDonald's Up 8.6%,' *Burger Business*, 14 March 2012. http://www.burgerbusiness.com/?p=9772

3 Conducted by Sponsorship Research International and cited in Schlosser, E., 2001. *Fast Food Nation.* New York: Houghton Mifflin Co, 276.

4 Emma Boyland, personal communication, 15 September 2012.

5 Helmer, J., 1992. 'Love on a bun: How McDonald's won the burger wars'. *Journal of Popular Culture*, 26 (2), 85–96.

6 Zhong, C-B. & DeVoe, S. E., 2010. 'You Are How You Eat: Fast Food and Impatience'. *Psychological Science* 21 (5), 619–22.

7 Schor, J., 2004. *Born to Buy.* New York: Scribner, 19–20.

8 Bruce, A. S., et al., 2010. 'Obese children show hyperactivation to food pictures in brain networks linked to motivation, reward and cognitive control'. *International Journal of Obesity*, 34 (10), 1494–500.

9 Harris, J. L., Bargh, J. A. &. Brownell, K. D., 2009. 'Priming Effects of Television Food Advertising on Eating Behaviour'. *Health Psychology*, 28 (4), 404–13.

10 Ibid.

11 McGinnis, J. M., Gottman, J. A. & Kraak, V. I., eds, 2006. *Food Marketing to Children and Youth: Threat or Opportunity?* Washington, DC: National Academies Press; Harris, J. L., et al., 2009. 'A crisis in the marketplace: how food marketing contributes to childhood obesity and what can be done'. *Annual Review of Public Health* 30 (2009), 211–25.

12 Kunkel, D., et al., 2004. 'Report of the APA task force on advertising and children'. American Psychological Association. www.apa.org/releases/childrenads.pdf. Total media spending increased from $197 million in 2008 to $264 million in 2011 http://www.cerealfacts.org/media/Cereal_FACTS_Report_Summary_2012_7.12.pdf

13 Page, R., et al., 2008. 'Targeting children in the cereal aisle: Promotional techniques and content features on ready-to-eat cereal product packaging'. *American Journal of Health Education*, 39 (5), 272–82.

14 Harris, J. L., Schwartz, M. B. & Brownell, K. D., 2012. *Cereal F.A.C.T.S. A spoonful of progress in a bowl full of unhealthy marketing to children.* Yale: Yale Rudd Center, 3.

15 Harris, J. L., Bargh, J. A. &. Brownell, K. D., 2009. 'Priming Effects of Television Food Advertising on Eating Behaviour', *Health Psychology*, 28 (4), 404–13.

16 Garretson, J. A. & Niedrich, R. W., 2004. 'Spokes-characters: creating character trust and positive brand attitudes'. *Journal of Advertising*, 33 (2), 25–36.

17 Lapierre, M. A., Vaala, S. E. & Linebarger, D. L., 2011. 'Influence of licensed spokes-characters and health cues on children's ratings of cereal taste'. *Archives of Pediatric and Adolescent Medicine*, 165 (3), 229–34.

18 Roberto, C. A., et al., 2010. 'Influence of licensed characters on children's taste and snack preferences'. *Pediatrics* 126 (1), 88–93.

19 Robinson et al., 'Effects of fast food branding'; Garretson, J. A. & Niedrich, R. W., 2004. 'Spokes-characters: creating character trust and positive brand attitudes'. *Journal of Advertising*, 33 (2), 25–36.

20 Musicus, A., Tal, A. & Wansink, B., 2014. 'Eyes in the Aisles: Why is Cap'n Crunch Looking Down at My Child?' *Environment & Behavior*. In print.

21 Ibid.

22 Ibid.

23 Forsythe, T., 2014. http://www.blog.generalmills.com/2014/04/response-to-absurd-cereal-study

24 Harris, 'A spoonful of progress'.

25 Ibid.

26 Bragg, M. A., et al., 2012. *Public Health Nutrition*, 16 (4), 738–42.

27 Bragg, M. A., et al., 2014. 'Athlete Endorsements in Food Marketing'. *Pediatrics*, 132 (5), 805–10.

28 Ibid.

29 Boyland, E. J., et al., 2013. 'Food Choice and Overconsumption: Effect of a Premium Sports Celebrity Endorser'. *The Journal of Pediatrics*, 163 (2), 339–43.

30 Emma Boyland's comments were made in a press release from the University of Liverpool published in March 2013. http://news.liv.ac.uk/wp-content/uploads/2013/03/crisps1.jpg

31 Haelle, T., 2014. 'Studies: Celebrity Endorsement Encourage Unhealthy Food, Drink'. The Chicago Bureau, 31 January 2014. http://www.chicago-bureau.org/celebrity-endorsements-encourage-unhealthy-food-drinks

32 Wansink, B. & Chandon, P., 2006. 'Can "low-fat" nutrition labels lead to obesity?' *Journal of Market Research*, 43 (4), 605–617; Andrews, J. C., Netemeyer, R. G. &

Burton, S., 1998. 'Consumer generalization of nutrient content claims in advertising'. *Journal of Marketing*, 62 (4), 62–75.

33 Bailey Dougherty, personal communication, 2012.

34 Sanburn, J., 2011. 'NYC Grocery Store Pipes in Artificial Food Smells', *Time Magazine Moneyland*, 20 July. http://moneyland.time.com/2011/07/20/nyc-grocery-store-pipes-in-artificial-food-smells

35 Dublino, J., 2012. 'McCain Foods to Launch Multi-Sensory Bus Shelter Ads in UK', *Scent Marketing Digest*, 7 February. http://scentmarketingdigest.com/2012/02/07/mccain-foods-to-launch-multi-sensory-busshelter-ads-in-uk

36 Dublino, J., 2012. 'Multi Sensory Dunkin' Donut Campaign Spikes Sales'. *Scent Marketing Digest*, 9 April 2012. http://scentmarketingdigest.com/2012/04/09/multi-sensory-dunkin-donut-campaign-spikessales

37 Dan Jones, personal communication, 2012.

38 Rideout, V., 2013. 'Zero to Eight: Children's Media Use in America 2013, a Common Sense Media Research Study', www.commonsensemedia.org

39 Smith, S., 2013. 'Tablets evolving into our transaction screen'. *Mobile Insider*, 29 January, Mediapost.com.

40 Fogg, B. J., 1997. 'Charismatic Computers: creating More Likeable and persuasive Interactive Technologies by Leverage in Principles from Social Psychology'. Doctoral dissertation, Stanford University, Stanford, CA.

41 Scarpello, L., 2012. 'The Retailer's Guide to So-Lo-Mo', from Monetate.com http://monetate.com/infographic/the-retailers-guide-to-solomo/#axzz2DWH4zQbG

42 http://www.cerealfacts.org/media/Cereal_FACTS_Report_Summary_2012_7.12.pdf

43 Harris, 'Spoonful of progress'.

44 Diaz, C. A., 2009. 'Goodby and B-Reel Enter the Asylum for the Sequel to Doritos Hotel 626'. *Creativity on Line*, 23 September. Retrieved from http://creativity-online.com/news/goodby-and-breel-step-into-the-asylum-for-the-sequel-to-doritos-hotel-626/139224, 20 March 2013.

45 Montgomery, K. & Chester, J., 2011. *Digital Food Marketing to Children and Adolescents. Problematic Practices and Policy Interventions*. Report from Public Health, Law & Policy (phlp). Washington DC: Robert Wood Johnson, 1–65.

46 Microsoft, 2009. 'Games Advertising Strikes An Emotional Chord With Consumers,' 15 June, www.emsense.com/press/game-advertising. php (viewed 11 April 2010). See also Microsoft Advertising, 'Doritos Xbox Live Arcade Game

Smashes Records', 21 June 2011. http:// advertising.microsoft.com/europe/doritos-xbox-live-arcade-game

47 Liikkanen, L. A., 2012. 'Involuntary music among normal populations and clinical cases', ACNR, 12 (4), 12–14.

48 Gulas, C. S. & Schewe, C. D., 1994. *Atmospheric Segmentation: Managing Store Image With Background Music, Enhancing Knowledge Development in Marketing*, Chicago IL: American Marketing Association, 325–30.

49 Milliman, R. E., 1982. 'Using background music to affect the behavior of supermarket shoppers'. *Journal of Marketing*, 46 (3), 86–91.

50 Blattner, M. M., Sumikawa, D. A. & Greenberg, R. M., 1989. 'Earcons and Icons: Their Structure and Common Design Principles', *Human-Computer Interaction*, 4 (1989), 11–14.

51 Fox, B., 2002. 'Audible Icons'. *New Scientist* 176 (2371), 18.

52 Watson, J. B., 1924. *Behaviourism*. New York: W.W. Norton.

Chapter 12 – Whatever Have They Done to Our Food?

1 *Forbes*, 8 June 2012. http://www.forbes.com/sites/csr/2012/08/06/choice-at-the-supermarket-is-our-food-system-the-perfect-oligopoly

2 Schwartz, B., 2004. *The Paradox of Choice*, New York: HarperCollins, 9–10.

3 Stull, D. & Broadway, M., 2012. 'Freedom of Choice at the Supermarket: Not'. CSRwire's blog *Talkback*. http://www.csrwire.com/blog/posts/454-freedom-of-choice-at-the-supermarket-not

4 Atkins, P. & Bowler, I., 2007. *Food in Society: economy, culture, geography.* London: Hodder Education, 101.

5 http://web.mit.edu/invent/iow/mccormick.html

6 Biello, S., 2009. 'Norman Borlaug: Wheat breeder who averted famine with a "Green Revolution"'. *Scientific American*. http://www.scientificamerican.com/blog/post/norman-borlaug-wheat-breeder-who-av-2009-09-14/?id=norman-borlaug-wheat-breeder-who-av-2009-09-14

7 http://www.theatlantic.com/magazine/archive/1997/01/forgotten-benefactor-of-humanity/306101

8 Olson, J. T., 2013. *Putting Meat on the Table: Industrial Farm Animal Production in America.* Baltimore, Maryland: Johns Hopkins.

9 Trewavas, A., 2002. Malthus foiled again and again. *Nature.* 418 (6898), 668–70.

10 http://www.theatlantic.com/magazine/archive/1997/01/forgotten-benefactor-of-humanity/306101/?single_page=true accessed 17 July 2014.

11 http://www.forbes.com/sites/rosspomeroy/2014/03/10/greens-vs-the-green-revolution/ accessed 17 July 2014.

12 http://www.theatlantic.com/magazine/archive/1997/01/forgotten-benefactor-of-humanity/306101/?single_page=true accessed 17 July 2014.

13 http://ec.europa.eu/research/agriculture/scar/pdf/scar_feg_ultimate_version.pdf, Sustainable food consumption and production in a resource-constrained world.

14 Atkins & Bowler, Food in Society.

15 Brownell, K. D. & Warner, K. E., 2009. 'The perils of ignoring history: Big Tobacco played dirty and millions died. How similar is Big Food?' *The Milbank Quarterly*, 87 (1), 259–94.

16 Stuckler, D. & Nestle, M., 2012. 'Big food, food systems, and global health'. *PLoS medicine*, 9 (6), e1001242.

17 Mayhew, C. & Quinlan, M., 2002. 'Fordism in the fast food industry: pervasive management control and occupational health and safety risks for young temporary workers'. *Sociology of Health & Illness*, 24 (3), 261–84.

18 https://www.gov.uk/government/uploads/system/uploads/attachment_data/file/315418/foodpocketbook-2013update-29may14.pdf

19 Tubiello, F. N., et al., 2014. 'Agriculture, Forestry, and Other Land Use Emissions by Sources and Removals by Sinks 1990–2011'. Analysis, accessed 14 September 2014, http://www.fao.org/docrep/019/i3671e/i3671e.pdf

20 http://www.oecd.org/env/indicators-modelling-outlooks/49910023.pdf

21 2014. 'Polder and Wiser'. *The Economist*, 23 August, pp. 61–2. http://www.economist.com/news/business/21613356-dutch-farmers-add-sustainability-their-enviable-productivity-polder-and-wiser

22 'The light fantastic: Indoor farming may be taking root'. *The Economist*, 17 May 2014. http://www.economist.com/news/science-and-technology/21602194-indoor-farming-may-be-taking-root-light-fantastic

23 Colangelo is cited from Gray, K., 2014. 'Server Farm', *Wired Magazine*, October, 123–31.

24 Abraham, J., 1991. *Food and development: the political economy of hunger and the modern diet*. London: Kogan Page; Weaver, C. M., et al., 2014. 'Processed foods: contributions to nutrition'. *American Journal of Clinical Nutrition*, 99 (6), 1525–42; Nestle, M., 2013. 'Conflicts of interest in nutrition societies'. *American Society of Nutrition*, http://www.foodpolitics.com/2013/11/conflicts-of-interest-in-nutrition-societies-american-society-of-nutrition, downloaded 8 August 2014.

25 Klein, S., 2010. 'Fatty food may cause cocaine-like addiction'. *CNN Health*. http://www.cnn.com/2010/HEALTH/03/28/fatty.foods.brain/ accessed 15 September 2014.

26 Weaver et al., 'Processed foods'.

27 Nestle, 'Conflicts of interest'.

28 Fiore, K., 2014. 'Nutritionists Pan ASN Processed Foods Statement', *MedPage Today*, 28 July; Stern, A., 2012. '"Pink slime" producer allows tour of plant to bolster image', South Sioux City. 29 March, 7:57 p.m. edition. http://www.reuters.com/article/2012/03/29/us-food-slime-idUSBRE82S1I520120329

29 Katz, D., 2014. 'Processing Messages About Processed Food'. *Huffington Post* http://www.huffingtonpost.com/david-katz-md/diet-and-nutrition_b_5644727 accessed 8 August 2014.

30 Stern, '"Pink slime" producer'.

31 Monteiro, C. A. & Cannon, G., 2012. 'The Impact of Transnational "Big Food" Companies on the South: A View from Brazil'. *Public Library of Science Medicine*, 9 (7), e1001252.

32 Moss quoted in Fassler, J., 2013. 'The Language of Junk Food Addiction: How to 'Read' a Potato Chip.' *The Atlantic*, April 30, 2013 (http://www.theatlantic.com/health/archive/2013/04/the-language-of-junk-food-addiction-how-to-read-a-potato-chip/275424/ accessed January 15, 2015).

33 Monteiro & Cannon, 'The Impact of Transnational "Big Food" Companies'.

34 Canella, D. S., et al., 2014. 'Ultra-Processed Food Products and Obesity in Brazilian Households, 2008–2009'. *Public Library of Science*, 9 (3), 1–9.

35 Moss, M., 2013. 'The Extraordinary Science of Addictive Junk Food'. *The New York Times Magazine*, 20 February.

36 http://www.preparedfoods.com/ext/resources/Special_Reports/Sweetening-the-Pot.pdf

37 McBride cited in Kessler, D., 2010. *The End of Overeating*. New York, Rodale.

38 Moskowitz, H., personal communication, 22 July 2014.

39 Chui, M., et al., 2012. 'The social economy, Unlocking value and productivity through social technologies. *The McKinsey Report*, July.

40 http://www.cbn.com/health/nutrition/reinke_label-lingo.aspx accessed 17 July.

41 http://www.globalresearch.ca/the-globalization-of-fast-food-behind-the-brand-mcdonald-s/25309

42 Chou, S., Grossman, M. & Saffer, H., 2004. 'An Economic Analysis of Adult

Obesity: Results from the Behavioral Risk Factor Surveillance System'. *Journal of Health Economics*, 23 (3), 565–87.

43 Ritzer, G., 2007. *The McDonaldization of Society*. London: Sage Publications.

44 Pan, A., Malik, V. S. & Hu, F. B., 2012. 'Exporting diabetes mellitus to Asia: the impact of Western-style fast food'. *Circulation*, 126 (2), 163–5.

45 White, J. S., 2008. 'Straight talk about high-fructose corn syrup: what it is and what it ain't'. *The American Journal of Clinical Nutrition*, 88 (6), 1716S–1721S; Dunn R., 2008. 'The Effect of Fast-Food Availability on Obesity: An Analysis by Gender, Race, and Residential Location'. *American Journal of Agricultural Economics*, 92 (4), 1139–64.

46 De Vogli, R., Kouvonen, A. & Gimeno, D., 2014. 'The influence of market deregulation on fast food consumption and body mass index: a cross-national time series analysis'. *Bulletin of the World Health Organization*, 92 (2), 99–107, 107A.

47 Ibid.

48 Grover, A., 2014. Special Rapporteur to health, 'Unhealthy foods, non communicable diseases, and the right to health'. http://www.unscn.org/files/Announcements/Other_announcements/A-HRC-26-31_en.pdf

49 De Vogli, et al., 'Influence of market deregulation'; Grover, 'Unhealthy foods'; Brownell & Warner, 'Perils of ignoring history'.

50 Wiist, W., 2011. 'The corporate play book, health, and democracy: the snack food and beverage industry's tactics in context'. In D. Stuckler & K. Siegel, eds. *Sick Societies: Responding to the global challenge of chronic disease*. Oxford: Oxford University Press.

51 Zadek, S., 2004. 'The Path to Corporate Responsibility'. *Harvard Business Review* 82 (December), 125–132.

52 Soda and Tobacco Industry Corporate Social Responsibility Campaigns: How Do They Compare? http://www.plosmedicine.org/article/info%3Adoi%2F10.1371%2Fjournal.pmed.1001241

53 Wiist, 'Corporate play book'.

54 Warner, M., 2010. 'The soda tax wars are back: Brace yourself'. *BNet*. Available: http://www.bnet.com/blog/food-industry/the-soda-tax-wars-are-back-brace-yourself/474 accessed 19 October 2011.

55 Ma, J., et al., 2014. 'Sweetened beverage consumption is associated with abdominal fat partitioning in healthy adults'. *Journal of Nutrition*, 144 (8), 1283–90.

56 Traill, B. W., 2006. 'Trends towards overweight in lower- and middle-income countries: some causes and economic policy options'. www document. http://www.fao.org/docrep/009/a0442e/a0442e0x.htm accessed September 2014.

57 Jacobsen, M. F., 2000. *Liquid Candy: how soft drinks are harming Americans' health*. Washington (DC): Center for Science in the Public Interest.

58 Basu, S., et al., 2013. 'Relationship of soft drink consumption to global overweight, obesity, and diabetes: a cross-national analysis of 75 countries'. *American Journal of Public Health*, 103 (11), 2071–7.

59 Hawkes, C., 2008. 'Dietary implications of supermarket development: a global perspective'. *Development Policy Review*, 26 (6), 657–92.

60 Dawson, J., 1995. *Food Retailing and the Consumer*. Glasgow: Blackie Academic and Professional; Dobson, P. W., Waterson, M. & Davies, S. W., 2003. 'The Patterns and Implications of Increasing Concentration in European Food Retailing'. *Journal of Agricultural Economics*, 54 (1), 111–25.

61 Heffernan, W. & Hendrickson, M., 2007. 'The Global Food System: A Research Agenda'. In: *Report to the Agribusiness Accountability Initiative Conference on Corporate Power in the Global Food System, High Leigh Conference Centre, Herts, UK*.

62 http://corporate.walmart.com/our-story/our-business/locations/accessed September 2014.

63 Hawkes, C., 2008. 'Dietary Implications of Supermarket Development: A Global Perspective'. *Development Policy Reviews*, 26 (6), 657–92.

64 USDA, 2014. Farmers Markets and Local Food Marketing. *National Count of Farmers Market Directory Listing Graph 1994–2014*. http://www.ams.usda.gov/AMSv1.0/ams.fetchTemplateData.do?template=TemplateS&leftNav=WholesaleandFarmersMarkets&page=WFMFarmersMarketGrowth&description=Farmers+Market+Growth accessed September 2014; Bardo, M. & Warwicker, M., 2012. 'Does Farmers' market food taste better?' *BBC News*. http://www.bbc.com/news/business-18522656 accessed September 2014.

65 Sexton, S., 2011. The Inefficiency of Local Food. *Freakonomics.com* http://freakonomics.com/2011/11/14/the-inefficiency-of-local-food/ accessed September 2014.

66 Ibid.

67 Frediani, K., 2013. 'Urban Food Production', presentation, http://www.youtube.com/watch?v=lK5qQCEDwa8 accessed September 2014.

68 Hawkes, C., 2007. 'Regulating and litigating in the public interest: regulating food marketing to young people worldwide: trends and policy drivers'. *American Journal of Public Health*, 97 (11), 1962–73.

Part Five – Losing It Wrong – Losing It Right

1 Fell, J. & Leitch, M., 2014. *Losing it Right: A Brutally Honest 3-Stage Program to Help You Get Fit and Lose Weight Without Losing Your Mind.* Toronto. Random House Canada.

Chapter 13 – Losing It Wrong: Why Fad Diets Fail

1 Freedhoff, Y., 2014. *The Diet Fix.* New York: Random House.

2 Park, M., 2010. 'Twinkie diet helps nutrition professor lose 27 pounds'. http://www.cnn.com/2010/HEALTH/11/08/twinkie.diet.professor/ accessed August 2014.

3 Ibid.

4 Answers.com. 2014. 'How Many Calories Per Gram of Macronutrient?'

5 Forbes, L., 2014. Interview with Dr Laura Forbes.

6 Ibid.

7 Smolin, L.A. & Grosvenor, M. B., 2013. 'Carbohydrates: Sugars, Starches, and Fiber'. *Nutrition: Science and Applications, 3rd ed.* Vol 3. New Jersey: Wiley, 114–58.

8 Harvard School of Public Health, 2014. 'The Nutrition Source, Carbohydrates'. http://www.hsph.harvard.edu/nutritionsource/carbohydrates/ accessed August 2014.

9 Andrews, R., 2014. 'All about nutrient timing: Does when you eat really matter?'. *Precision Nutrition.* http://www.precisionnutrition.com/all-about-nutrient-timing

10 Davis, W., 2011. *Wheat Belly.* New York: Rodale; Perlmutter, D. & Loberg, K., 2013. *Grain Brain: The Surprising Truth about Wheat.* New York: Little Brown.

11 Gulli, C., 2013. 'The dangers of going gluten-free'. *Maclean's Magazine.* http://www.macleans.ca/society/life/gone-gluten-free/; Wheat Belly Lifestyles Institute. Wheat Belly FAQs, 2014. http://www.wheatbelly.com/articles/WBFAQs accessed 10 September 2014.

12 NHS Choices, 2013. 'Diagnosing coeliac disease'. http://www.nhs.uk/conditions/Coeliac-disease/Pages/Introduction.aspx accessed 20 August 2014.

13 Ibid.

14 Gulli, 'The dangers of going gluten-free'.

15 Facts, P., 2012. *Gluten-Free Foods and Beverages in the U.S., 4th Edition.* http://www.packagedfacts.com/Gluten-Free-Foods-7144767/2012

16 Ibid.

17 Ratner, A., 2011. 'Gluten Free Rice Krispies?' http://www.glutenfreeliving. com/2011/02/22/gluten-free-rice-krispies/ accessed 10 September 2014.

18 Gulli, 'The dangers of going gluten-free'.

19 Maguire, T. & Haslam, D., 2010. *The Obesity Epidemic and Its Management*. London: Pharmaceutical Press.

20 Forbes, interview.

21 Ibid.

22 Westerterp, K. R., Wilson, S. A. & Rolland, V., 1999. 'Diet induced thermogenesis measured over 24h in a respiration chamber: effect of diet composition'. *International Journal of Obesity and Related Metabolic Disorders*, 23 (3), 287–92.

23 Mikkelsen, P. B., Toubro, S. & Astrup, A., 2000. 'Effect of fat-reduced diets on 24-h energy expenditure: comparisons between animal protein, vegetable protein, and carbohydrate'. *The American Journal of Clinical Nutrition*, 72 (5), 1135–41; Robinson, S. M., et al., 1990. 'Protein turnover and thermogenesis in response to high-protein and high-carbohydrate feeding in men'. *The American Journal of Clinical Nutrition*, 52 (1), 72–80.

24 McNay, D. E. & Speakman, J. R., 2012. 'High fat diet causes rebound weight gain'. *Molecular Metabolism*. 2 (2), 103–8; St Jeor, S. T., et al., 2001. 'Dietary protein and weight reduction: a statement for healthcare professionals from the Nutrition Committee of the Council on Nutrition, Physical Activity, and Metabolism of the American Heart Association'. *Circulation*, 104 (15), 1869–74.

25 Forbes, interview.

26 Freedhoff, Y., 2014. Personal communication, Interview with Yoni Freedhoff, 17 July 2014.

Chapter 14 – Losing it Right:
Achieving Sustainable Weight Loss

1 Feynman, R., with Leighton, R., 1992. *Surely You're Joking, Mr. Feynman!* London: Random House, 342.

2 Wadden, T. A., et al., 2011. 'Four-year weight losses in the Look AHEAD study: factors associated with long-term success'. *Obesity*, 19 (10), 1987–98.

3 Wansink, B., Painter, J. E. & North, J., 2005. 'Bottomless bowls: why visual cues of portion size may influence intake'. *Obesity research*, 13 (1), 93–100.

4 Wansink, B., van Ittersum, K. & Painter, J. E., 2006. 'Ice cream illusions bowls, spoons, and self-served portion sizes'. *American Journal of Preventive Medicine*, 31 (3), 240–3.

5 Wansink, B. & Ittersum, K., 2003. 'Bottoms Up! The Influence of Elongation on Pouring and Consumption'. *Journal of Consumer Research*, 30 (December), 455–63.

6 Wansink, B., 2006. *Mindless Eating – Why We Eat More Than We Think*. New York: Bantam-Dell, 2006.

7 Ibid; Wansink, B. & Linder, L. R., 2003. 'Interactions between forms of fat consumption and restaurant bread consumption'. *International Journal of Obesity and Related Metabolic Disorders*, 27 (7), 866–8.

8 Wansink, B. & Sobal, J., 2007. 'Mindless eating: the 200 daily food decisions we overlook'. *Environment & Behaviour*, 39 (1), 106–123.

9 'Do low-fat foods make us fat? Mindless eating leads us to eat 28-45% more calories when foods are "low-fat"' [press release]. http://www.eurekalert.org/pub_releases/2006-12/cfb-dlf120806.php2006

10 Wansink, B. & Chandon, P., 2006. 'Can "Low-Fat" Nutrition Labels Lead to Obesity?' *Journal of Marketing Research*. 43 (4), 605–17.

11 Ibid.

12 Ibid.

13 Ibid.

14 Ferrerò, G., 1894. 'L'inertie mentale et la loi du moindre effort'. *Revue Philosophique de la France et de l'Étranger*, 37 (January–June), 169–182; Tolman, E. C., 1932. *Purposive behavior in animals and men*. Berkeley, CA: University of California Press.

15 Rozin, P., et al., 2011. 'Nudge to nobesity: Minor changes in accessibility decrease food intake'. *Judgement and Decision Making*. 6 (4), 323–32.

16 Gallant, A. R., Lundgren, J. & Drapeau, V., 2012. 'The night-eating syndrome and obesity'. *Obesity Reviews*, 13 (6), 528–36.

17 Frecka, J. M. & Mattes, R. D., 2008. 'Possible entrainment of ghrelin to habitual meal patterns in humans'. *American Journal of Physiology. Gastrointestinal and liver physiology*, 294 (3), G699–707.

18 Davis, A., 1971. *Let's Eat Right to Keep Fit*. New York: Harcourt.

19 Sjoberg, A. et al., 2003. 'Meal pattern, food choice, nutrient intake and lifestyle factors in The Goteborg Adolescence Study'. *European Journal of Clinical Nutrition*, 57 (12), 1569–78; Savige, G., et al., 2007. 'Snacking behaviours of adolescents and their association with skipping meals'. *The International Journal of Behavioral Nutrition and Physical Activity*, 4 (2007), 36.

20 Westerterp-Plantenga, M. S., et al., 2009. 'Dietary protein, weight loss, and weight maintenance'. *Annual Review of Nutrition*, 29 (2009), 21–41.

21 Lejeune, M. P., et al., 2006. 'Ghrelin and glucagon-like peptide 1 concentrations, 24-h satiety, and energy and substrate metabolism during a high-protein diet and measured in a respiration chamber'. *The American Journal of Clinical Nutrition*, 83 (1), 89–94; Luhovyy, B. L., Akhavan, T. & Anderson, G. H., 2007. 'Whey proteins in the regulation of food intake and satiety'. *Journal of the American College of Nutrition*, 26 (6), 704S–712S.

22 *Secret Eaters*, 2014, London, season 1, episode 4.

23 Bell, R. & Pliner, P.L., 2003. 'Time to eat: the relationship between the number of people eating and meal duration in three lunch settings'. *Appetite*, 41 (2), 215–18.

24 Herman, C. P., Roth, D. A. & Polivy, J., 2003. 'Effects of the presence of others on food intake: a normative interpretation'. *Psychological Bulletin*, 129 (6), 873–86; Wansink, B., 2004. 'Environmental factors that increase the food intake and consumption volume of unknowing consumers'. *Annual Review of Nutrition*, 24 (2004), 455–79.

25 My Fitness Pal, 2014. Calories in a Large Glass of Wine.

26 Wansink, B., Payne, C. R. & Chandon, P., 2007. 'Internal and external cues of meal cessation: the French paradox redux?' *Obesity*, 15 (12), 2920–4.

27 Wing, R. R. & Hill, J. O., 2001. Successful weight loss maintenance. *Annual Review of Nutrition*, 21 (2001), 323–41; Wing, R. R. & Phelan, S., 2005. 'Long-term weight loss maintenance'. *The American Journal of Clinical Nutrition*, 82 (1 Suppl.), 222S–225S.

28 Hoevenaar-Blom, M. P., et al., 2013. 'Sufficient sleep duration contributes to lower cardiovascular disease risk in addition to four traditional lifestyle factors: the MORGEN study'. *European Journal of Preventive Cardiology*, 2013. Epub Ahead of print at time of writing *Fat Planet*.

29 Fell, J. & Leitch, M., 2014. *Lose it Right: A Brutally Honest 3-Stage Program to Help You Get Fit and Lose Weight Without Losing Your Mind*. Toronto, Canada: Random House.

Chapter 15 – Planning for a Slimmer Planet

1 Cohen, D., personal communication, 2014.

2 US Department of Agriculture, Economic Research Service, Food and Rural Economics Division, 2012. 'America's Eating Habits: Changes and Consequences'. http://www.ers.usda.gov/media/92736/aib750_1_.pdf accessed September 2014.

3 US Department of Agriculture Office Communications Agriculture Fact Book 2001-2002, in Chapter 2: Profiling Food Consumption in America. Washington DC. http://www.usda.gov/factbook/chapter2.pdf

4 Piggot, J., 2014. Court ruling. Sugary Drinks Portion Cap Rule. https://www.nycourts.gov/ctapps/Decisions/2014/Jun14/134Opn14-Decision.pdf

5 Cohen, personal communication, 2014.

6 Phillip, A., 2014. 'A Company Pushing Bogus Diet Pills Touted by Dr. Oz Settles With the FTC. Will the Medical World Weigh In?' *Washington Post.* http://www.washingtonpost.com/news/to-your-health/wp/2014/09/09/the-ftc-fined-a-company-pushing-diet-pills-touted-by-dr-oz-will-the-medical-world-weigh-in/ accessed 5 September 2014.

7 Leslie, D., 2014. 'The Hollywood Medical Reporter – the Land of Oz'. *Brainblogger.com.* http://brainblogger.com/2014/09/07/the-hollywood-medical-reporter-the-land-of-oz/ accessed 10 September 2014.

8 Gavura, S., 2014. 'Dr. Oz and the Terrible, Horrible, No Good, Very Bad Day'. *Science Based Medicine.* http://www.sciencebasedmedicine.org/tag/dr-oz/ accessed 10 September 2014.

9 Cifani, C., et al., 2009. 'A preclinical model of binge eating elicited by yo-yo dieting and stressful exposure to food: effect of sibutramine, fluoxetine, topiramate, and midazolam'. *Psychopharmacology,* 204 (1), 113–25; Haiken, M., 2014. 'FDA Postpones Approval Of Contrave: What's The Future Of Weight Loss Drugs?' *Forbes.* http://www.forbes.com/sites/melaniehaiken/2014/06/11/new-weight-loss-drug-contrave-postponed-by-fda/ accessed 10 August 2014; Verpeut, J. L. & Bello, N. T., 2014. 'Drug safety evaluation of naltrexone/bupropion for the treatment of obesity'. *Expert Opinions in Drug Safety,* 13 (6), 831–41.

10 Frances, A., 2013. *Saving Normal: An Insider's Revolt Against Out-Of-Control Psychiatric Diagnosis, Dsm-5, Big Pharma, and the Medicalization of Ordinary Life.* New York: HarperCollins.

11 Ibid.

12 Kraak, V. I. & Story, M., 2010. 'A public health perspective on healthy lifestyles and public–private partnerships for global childhood obesity prevention'. *Journal of the American Dietetic Association,* 110 (2), 192–200.

13 Hawkes, C. & Buse, K., 2011. 'Public health sector and food industry interaction: it's time to clarify the term "partnership" and be honest about underlying interests'. *European Journal of Public Health,* 21 (4), 400–1.

14 Obama, M., 2014. Let's Move! [www website] www.letsmove.org

15 World Health Organization, '2008–2013 Action Plan for the Global Strategy for the Prevention and Control of Noncommunicable Diseases'. http://www.who.int/nmh/publications/ncd_action_plan_en.pdf accessed September 2014.

16 Borys, J. M., et al., 2012. 'EPODE approach for childhood obesity prevention: methods, progress and international development'. *Obesity Reviews,* 13 (4), 299–315.

17 Diet and Health Research Industry Club (DRINC) http://www.bbsrc.ac.uk/drinc

18 https://responsibilitydeal.dh.gov.uk/pledges/

19 Hawkes & Buse, 'Public health sector'.

20 Lang, T., Rayner, G. & Laelin, E., 2006. *The Food Industry, Diet, Physical Activity and Health: A Review of Reported Commitments and Practice of 25 of the World's Largest Food Companies.* London: City University; Hawkes, C., Harris, J. L., 2011. 'An analysis of the content of food industry pledges on marketing to children'. *Public Health Nutrition*, 14 (8), 1403–14.

21 Wootan, M., Batada, A. & Balkus, O., 2010. 'Food Marketing Report Card: An Analysis of Food and Entertainment Company Policies to Self-Regulate Food and Beverage Marketing to Children, 2010'. http://cspinet.org/new/pdf/marketingreportcard.pdf accessed September 2014; King, L., et al., 2013. 'Building the case for independent monitoring of food advertising on Australian television'. *Public Health Nutrition*, 16 (12), 2249–54; Sacks, G., et al., 2013. 'A proposed approach to monitor private-sector policies and practices related to food environments, obesity and non-communicable disease prevention'. *Obesity Reviews*, 14 (Suppl 1), 38–48; Yach, D., et al., 2004. 'The global burden of chronic diseases: overcoming impediments to prevention and control'. *Journal of the American Medical Association*, 291 (21), 2616–22.

22 King, L., et al., 2013 'Building the case for independent monitoring of food advertising on Australian television'. *Public Health Nutrition*, 16(12), 2249-54.

23 Access to Nutrition Index, 2013. http://www.accesstonutrition.org accessed September 2014.

24 Swinburn, B., et al., 2013. 'INFORMAS (International Network for Food and Obesity/non-communicable diseases Research, Monitoring and Action Support): overview and key principles'. *Obesity Reviews*, 14 (Suppl 1), 1–12.

25 Womack, S., 2006. 'The Children Eating 9 L of Crisp Oil a Year'. *Daily Telegraph.* http://www.telegraph.co.uk/news/uknews/1529490/The-children-eating-nine-litres-of-crisp-oil-a-year.html accessed 8 August 2014.

26 Niederdeppe, J., Farrelly, M. C. & and Haviland M. L., 2004. 'Confirming "truth": more evidence of a successful tobacco countermarketing campaign in Florida'. *American Journal of Public Health*, 94 (2), 255–7; Hicks, J., 2001. 'The strategy behind Florida's "truth" campaign'. *Tobacco Control*, 10 (Spring), 3–5.

27 Kelley, A. E., Schiltz, C. A. & Landry, C. F., 2005. 'Neural systems recruited by drug- and food-related cues: studies of gene activation in corticolimbic regions'. *Physiology & Behavior*, 86 (1–2), 11-4; Erlanson-Albertsson, C., 2005. 'Sugar triggers our reward-system. Sweets release opiates which stimulate the appetite for sucrose – insulin can depress it'. *Lakartidningen*, 102 (21), 1620–2, 1625, 1627.

28 Kersley, R. & O'Sullivan, M., 2013. *Sugar Consumption at a Crossroads.* Switzerland: Credit Suisse Research Institute.

29 Cohen, personal communication, 21 August 2014.

30 Cohen, D., 2014. *A Big Fat Crisis: The Hidden Forces Behind the Obesity Epidemic – and How We Can End It.* New York, NY: Nation Books, 117.

31 Schultz, E. J., 2012. 'FTC Tallies Kid-Targeted Food Marketing at $1.79 Billion'. http://adage.com/article/news/ftc-tallies-kid-targeted-food-marketing-1-7-billion/238904/ accessed 10 August 2014.

32 Castellanos, E. H., et al., 2009. 'Obese adults have visual attention bias for food cue images: evidence for altered reward system function'. *International Journal of Obesity*, 33 (9), 1063–73; Jastreboff, A. M., et al., 2013. 'Neural correlates of stress- and food cue-induced food craving in obesity: association with insulin levels'. *Diabetes Care*, 36 (2), 394–402.

33 Chorley, M., 2013. 'Junk food at checkouts under fire as new health minister says she wants to stop parents being pestered by children'. *Daily Mail* online. http://www.dailymail.co.uk/news/article-2479785/Junk-food-WILL-banned-checkouts.html accessed 10 August 2014.

34 Poulter, S., 2014. 'Sweets at the checkout will stay as stores rebel over ban: Supermarkets resist call from health campaigners to remove temptation from the tills'. *Daily Mail* online. http://www.dailymail.co.uk/news/article-2608182/Sweets-checkout-stay-stores-rebel-ban-Supermarkets-resist-call-health-campaigners.html accessed 10 August 2014; Smithers, R., 2014. 'Tesco bans sweets from checkouts in all stores'. *The Guardian.* http://www.theguardian.com/business/2014/may/22/tesco-bans-sweets-from-checkouts-all-stores accessed 11 August 2014.

35 Cardello, H., Leitch, M.A. , 2012. Better for You Foods: It's Just Good Business (Hudson Institute, 2012). http://www.hudson.org/files/documents/BFY%20 Foods%20Executive%20Summary.pdf

36 Begley, S., 2014. Reuters. 'Food, beverage companies slash calories in obesity fight'.(http://www.reuters.com/article/2014/01/09/us-calories-idUSBREA0805F20140109,
accessed October 10, 2014).

37 Tarantola, A., 2014. 'Chicago's Huge Vertical Farm Glows Under Countless LED Suns'. http://gizmodo.com/chicagos-huge-vertical-farm-farm-glows-under-countless-1575275486

38 Philips News Centre, 2013. 'Philips & Green Sense Farms usher in new era of indoor farming with LED 'light recipes' that help optimize crop yield and quality'. http://www.newscenter.philips.com/us_en/standard/news/press/2014/20140509-philips-greensense-indoor-farming.wpd#.VLg8UmTF9hw

39 Delano, A., 2013. 'Food and Beverage Companies Surpass 2015 Goal of Reducing Calories in the US Three Years Ahead of Schedule'. http://www.healthyweightcommit. org/news/food_and_beverage_companies_surpass_2015_goal_of_reducing_ calories_in_the_u/ 2013 accessed 10 August 2014; 'Food, beverage companies slash calories in obesity fight'. *Daily News*, 13 January 2014. http://www.ift.org/food-technology/daily-news/2014/january/13/ food-beverage-companies-slash-calories-in-obesity-fight.aspx accessed August 2014

40 Ultragrain Whole Grain Nutrition webpage, W.F.A. http://www.ultragrain.com

41 Cardello, H., 2014. 'Why Big Food Belongs in the School Lunchroom'. *Forbes* 2014, http://www.forbes.com/sites/forbesleadershipforum/2014/09/04/why-big-food-belongs-in-the-school-lunchroom/ accessed 10 September 2014.

42 Yuan, L., et al., 2013. 'Metabolic mediators of the effect of body-mass index, overweight, and obesity on coronary heart disease and stroke: a pooled analysis of 97 prospective cohorts with 1.8 million participants'. *The Lancet*, 383 (November 2013), 970–83; Flegal, K. M., Kit, B. K., Graubard, B. I., 2013. 'Association of all-cause mortality with overweight and obesity using standard body mass index categories: a systematic review and meta-analysis'. *Journal of the American Medical Association*, 309 (1), 71–82.

43 Dobbs, R., et al., 2014. *Overcoming Obesity: An Initial Economic Analysis.* London: McKinsey Global Institute.

Index